T0213681

Molecular Dynamics and Complexity in Catalysis and Biocatalysis

Marco Piumetti

Molecular Dynamics and Complexity in Catalysis and Biocatalysis

 Springer

Marco Piumetti
Department of Applied Science
and Technology
Politecnico di Torino
Turin, Italy

With Contributions by
Andrés Illanes
School of Biochemical Engineering
Pontificial Catholic University of Valpa
Valparaíso, Chile

Nik Lygeros
Laboratoire de Génie des Procédés
Catalytiques
Villeurbanne, France

ISBN 978-3-030-88502-1   ISBN 978-3-030-88500-7   (eBook)
https://doi.org/10.1007/978-3-030-88500-7

This Springer imprint is published by the registered company Springer Nature Switzerland AG
The registered company address is: Gewerbestrasse 11, 6330 Cham, Switzerland

Preface

In recent decades, there has been great progress in research related to the science of catalysis and related applications due to advances in synthesis procedures, physico-chemical characterizations and testing conditions. The continuous growth in computer-related technologies has played a key role in the modelling and simulation of catalytic systems, thereby stimulating the development of novel catalytic materials that are effective under specific operating conditions. In this context, studies investigating active sites in different catalytic systems have been conducted using complementary approaches. Catalytic reactions are all based on the same principle, namely a reduction in the activation energy associated with the breaking and formation of new chemical bonds at active sites. Despite this common feature, heterogeneous, homogeneous and enzymatic catalyses represent distinct research areas that have been developed independently with modest reciprocal influence.

This monograph presents a concise comparison of the three main catalytic systems, namely enzymes and homogeneous and heterogeneous catalysts, outlining their catalytic properties and peculiarities. Moreover, a brief introduction to the science of catalysis unifying different catalytic systems into a single, conceptually coherent structure is presented. In fact, molecular dynamics and complexity can be observed in both catalysts and biocatalysts, with many similarities in both their structural configuration and operational mechanisms. The presence of active sites is a common feature of these catalytic systems, and studying their behaviour is critical for predicting catalytic results and designing effective catalytic materials.

Overall, this book does not provide a comprehensive review of molecular dynamics, reaction mechanisms and kinetics in catalysis and biocatalysis but rather a subjective picture and a summary of the lectures given during the Ph.D. course *Catalysis and Biocatalysis* taught by Marco Piumetti at the Pontifical Catholic University of Valparaíso (2019) and the Politecnico di Torino (2020). For convenience, the text is divided into five parts. The first part reviews the structure of proteins and their main features. Correct protein folding is crucial for protein functioning, especially for enzymes. The second part describes the configuration of enzymes, their kinetics and some factors that influence enzyme activity. The third part provides an introduction to molecular catalysis, including kinetics and the fundamental concepts of heterogeneous catalysis. Part four aims to illustrate the complex behaviour of active sites in heterogeneous, homogeneous and enzymatic catalysis. The final part addresses the study of complexity in catalytic processes, with a focus on advanced modelling approaches. Overall, I hope that readers will find this book pedagogically and otherwise helpful. Readers are advised to keep in mind that statements, data, figures, procedural details or other items can inadvertently include mistakes. Any comments can be sent to the authors.

I am greatly indebted to my coworkers who collaborated with me in the preparation of this work. In particular, I am profoundly grateful to Clarissa Cocuzza and Enrico Sartoretti for their excellent support. I thank also Dr. G. Demaria (Editor of Chemistry Today) and M. Ruffino (President of CLUT-Politecnico) who gave me the permission to reuse sections of my published works to prepare Chaps. 4 and 5. Moreover, many colleagues offered advice during the preparation of this book, and it is a pleasure to express my gratitude to them for their help and assistance.

Turin, Italy   Marco Piumetti
June 2021

About This Book

In this book, a concise comparison of catalytic and biocatalytic systems is provided outlining their catalytic properties and peculiarities. Moreover, this book presents a brief introduction to the science of catalysis and attempts to unify different catalytic systems into a single, conceptually coherent structure. In fact, molecular dynamics and complexity may occur in both catalysts and biocatalysts, with many similarities in both their structural configuration and operational mechanisms. However, this book is not a comprehensive review of molecular dynamics, reaction mechanisms and kinetics but rather provides a subjective picture and a summary of the lectures given during the Ph.D. course *Catalysis and Biocatalysis* taught by Marco Piumetti at the Pontifical Catholic University of Valparaíso, Chile (2019), and the Politecnico di Torino, Italy (2020). Thus, this book aims to be a didactic tool for designing introductory and more advanced conceptual courses at the undergraduate and graduate chemistry and engineering levels.

Introduction

Living organisms are made up of cells that carry out the life processes of nutrition, movement, growth, reproduction, respiration, sensitivity and excretion. All these processes can be properly performed due to the hierarchical, multiscale organization among biological structures, i.e. proteins, DNA, nuclei, cells, tissues and organs. Each single cell has a well-defined organization in which its components may interact with each other within a thermodynamically open system. In this almost perfect structural order, enzymes, which are biological catalysts, allow metabolic processes to occur at temperatures compatible with life. Each living process occurs due to enzymes. For instance, bacteria have a cell wall made of peptidoglycan, and inside the cell wall, the cell membrane contains several enzymes able to carry out physiological processes at a low temperature. Frequently, multiple enzymes are required to catalyse biochemical reactions; for example, the conversion of glucose to alcohol involves twelve consecutive steps and different enzymes. Thus, the metabolism of a single cell involves many enzyme-catalysed reactions and is structured according to numerous connected metabolic pathways. The latter typically occurs with sequential reactions, namely the product of the first reaction becomes the reagent in the second reaction, etc. Moreover, metabolic cycles can be formed when sets of reactions are arranged in loops (e.g. Krebs and Calvin cycles).

In multistep processes (e.g. glycolysis), some reactions occur under steady-state conditions, and the reaction rates increase or decrease as a function of the substrate concentration. Other reactions are far from equilibrium and are typically points of regulation of the overall pathway. In this highly complex network of reactions, the efficiency of an organism's metabolism is guaranteed by the presence of enzymes able to act as catalysts and regulators; enzymes not only establish interactions with the substrate, thereby lowering the activation energy of reactions, but also recognize which substrate to bind due to the stereospecificity of catalytic sites. Moreover, enzymes present flexible structures, and hence, their 3-D cavities can be easily adaptable to substrates and operating conditions. Consequently, cells can regulate their metabolic pathways via sophisticated mechanisms over a wide timescale by either modifying the activity of enzymes or changing their amount.

Compared to man-made inorganic catalysts, enzymes in living organisms are highly efficient biocatalysts. The rate improvements caused by such enzymes are in the range of 5–17 orders of magnitude [1]. However, for a biocatalyst to be effective in an industrial context, it must be subjected to improvement and optimization to ensure high performances over long periods (long-term stability) under process conditions. Therefore, several technological approaches have been proposed to design enzyme-like systems.

Enzymes are proteins, namely chains of amino acids with 3-D structures, which are mainly determined by chemical interactions (e.g. hydrogen bonds, van der Waals interactions, disulphide bridges, etc.). Correct protein folding, which is driven by the minimization of Gibbs' free energy of the molecule, is an essential requirement for enzyme functioning. In fact, enzymes present active sites that are effective in binding the substrate (binding sites) and then performing a catalytic action (catalytic sites). The active site is often composed of a few amino acid residues, although the enzyme molecule may contain hundreds of such residues. The entire enzyme structure plays a role since the whole sequence supports the active site and guarantees its correct 3-D orientation. A similar situation can be observed in heterogeneous catalysts, such as platinum group metals (PGMs) deposited onto metal oxide supports. The latter can be used for several catalytic applications, including pollutant abatement, hydrogenation and reforming reactions. Similar to enzymes, metal oxide supports often have a structural function, allowing the proper dispersion and accessibility of PGM clusters. Frequently, the presence of multiple components influences solid-state chemistry since active phases (and promoters) may interact with each other. Therefore, the establishment of strong metal–support interactions (SMSI) can modify the electronic properties of active centres, further improving the catalytic activity and stability of PGMs [2]. Moreover, support may provide binding sites for reactants in the proximity of the interface with the PGM clusters, analogous to binding sites close to the catalytic sites of enzymes.

It is well known that catalytic support may participate in a reaction. For example, the interrelation between ceria (CeO_2) and PGMs can be so strong that the formation of a homogeneous phase containing both species is even possible. Indeed, the formation of a homogeneous ceria palladium layer has been achieved at the surface of ceria nanostructures after ball milling (to promote tight contact) with Pd nanoparticles. Such a great degree of mixing can be obtained due to the high affinity between the components with significant modifications in the electronic properties of the involved atoms.

In this scenario, it is fundamental to know the nature of active sites involved in catalytic and biocatalytic processes. In heterogeneous catalysis, the concept of an "active site" was introduced by Taylor in 1925 as follows: "*A surface may be regarded as composed of atoms in varied degrees of saturation by neighboring metal atoms. The varying degree of saturation of the catalyst atoms also involves a varying catalytic capacity of the surface atoms. There will be all extremes between the case in which all the atoms in the surface are active and that in which relatively few are active*" [3]. This definition is suitable for continuous solid catalysts (i.e. metals, alloys and many halides, oxides and sulphides), but it appears less appropriate for describing active sites that are spatially distributed inside open-structure solids. The single-site heterogeneous catalysts (SSHCs) introduced by Sir John

Meurig Thomas are characteristic examples of catalytic materials in which the active centres (e.g. metals), which consist of one or more atoms, are spatially isolated from one another and are uniformly distributed over high-area supports (e.g. mesoporous silicas), thereby forming metal-oxo entities. Single atoms may be anchored to a surface structural defect at an oxide support, i.e. magnesia (MgO). Each active site is spatially isolated from the other sites and exhibits the same interaction energy with reactant molecules [4]. Thus, SSHCs can be considered inorganic analogues of enzymes. Nevertheless, enzymes can behave as both heterogeneous and homogeneous catalysts as described in the following sections of this book. Furthermore, the kinetics and reaction mechanisms provide evidence that heterogeneous catalytic reactions may have a counterpart in enzyme kinetics. For example, many oxidation reactions occur via redox-type mechanisms (or Mars–van Krevelen mechanism) over metal oxide catalysts as follows: the reactant molecule is directly oxidized by the oxide lattice oxygen, and an oxygen vacancy is formed accompanied by a simultaneous reduction in the catalyst surface. Subsequently, the lattice oxygen is replenished by a reduction in gaseous oxygen. Similar redox-type mechanisms can be observed in enzymes acting via the so-called ping-pong mechanism (i.e. the oxidation of glucose by molecular oxygen into hydrogen peroxide and gluconolactone by glucose oxidase) and some homogeneous catalysts, i.e. oxidation of ethylene to acetaldehyde in the presence of $PdCl_2$ in the solution as a catalyst (Wacker process). The existence of similarities in the kinetics and reaction mechanisms of different catalytic systems suggests that active sites may act similarly to carry out chemical reactions. In fact, the active sites of catalysts and biocatalysts generate electric fields that are able to catalyse reactions.

References

1. D. Nelson, M.M. Cox, *Lehninger Principles of Biochemistry*, 7th edn. (Macmillan, 2017), pp. 75–105
2. M. Piumetti, S. Bensaid, T. Andana, N. Russo, R. Pirone, D. Fino, Appl. Catal. B, **205**, 455–468 (2017)
3. H.S. Taylor Proc. R. Soc. Lond. A, **108**, 105–111 (1925)
4. J.M. Thomas, R. Raja, D.W. Lewis, Angew. Chem. Int. Ed. **44**, 6456–6482 (2005)

Contents

About the Author

Marco Piumetti received his European Ph.D. in Materials Science and Technology at the Politecnico di Torino in 2010. He trained at several academic institutions, including the Laboratoire de réactivité de surface—Sorbonne Université and Fritz Haber Institute. He is an appointed professor teaching the courses *Industrial Biochemistry* and *Introduction to Sustainable Engineering* offered at the Politecnico di Torino. His research activities are currently conducted with the Catalytic Reaction Engineering for Sustainable Engineering (CREST) group of the Department of Applied Science and Technology (DISAT) at the Politecnico di Torino and concern the fields of catalysis, surface science, nanomaterials, biocatalysis and biotech applications.

Structure of Proteins

1.1 Introduction

Proteins and peptides are chemically similar as both are composed of amino acid residues held together by covalent bonds. These molecules are fundamental for the physiological functions of organisms and mediate virtually all processes that occur in cells, e.g., catalysing metabolic reactions, responding to signals transmitted from the extracellular environment, giving cells their shapes, and transporting molecules. In living organisms, proteins are formed during translation by cell ribosomes after the transcription of DNA to mRNA in the cell nucleus. In fact, ribosomes are protein factories; the genetic instructions encoded in DNA are transcribed into mRNA and then translated from mRNA into proteins using tRNA and ribosomes. Thus, translation deciphers the information in the genetic code to produce functional proteins. The whole process is called gene expression and represents the central dogma of molecular biology (Fig. 1.1). However, some organisms may also form peptides by nonribosomal peptide synthesis, which often uses amino acids other than the 20 amino acids encoded by the genetic code.

Peptides are smaller and structurally much simpler than proteins. In fact, peptides are typically defined as molecules with 2–50 amino acids, whereas polypeptides (proteins) are macromolecules comprising 50 or more amino acids and, thus, have higher molecular weights.

From a structural perspective, peptides may exhibit some secondary structures (often the structure is only metastable), but typically, there are no highly ordered structures.

The human body may generate tens of thousands of different proteins, which are encoded by approximately 20,000 genes and serve a multitude of physiological purposes [1]. The molecules in our immune systems are proteins, and many

Electronic Supplementary Material The online version of this chapter (https://doi.org/10.1007/ 978-3-030-88500-7_1) contains supplementary material, which is available to authorized users.

M. Piumetti, *Molecular Dynamics and Complexity in Catalysis and Biocatalysis*,
https://doi.org/10.1007/978-3-030-88500-7_1

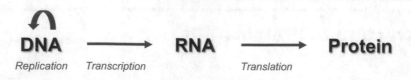

Fig. 1.1 Schematization of the steps of gene expression (the central dogma of molecular biology)

diseases result from protein misfolding causing proteins to lose functionality. Moreover, it is known that the accumulation of inactive aggregates of unfolded proteins in cells may cause diseases; more than 20 different diseases, including Alzheimer's disease, Parkinson's disease, Huntington's disease and Down's syndrome, are caused at least partially by abnormal protein aggregation.

Many cancers may result from mutations in proteins. It has been estimated that approximately 50% of human cancers are due to changes in the tumour protein p53 that compromise its structural stability. The TP53 *protein plays pivotal roles* in controlling cell division and cell death. An abnormal TP53 structure provides different information, which may cause cells to multiply uncontrollably and become cancerous. Therefore, the exact 3-D structure of a protein (native conformation) is crucial for its biological function. To study the activity of proteins, it is often necessary to determine their three-dimensional (3-D) structure via complementary techniques, such as X-ray crystallography, nuclear magnetic resonance (NMR) spectroscopy, and electron microscopy (vide infra).

Overall, the structure of proteins is typically described by the following conceptual hierarchy:

The primary structure refers to the sequence of amino acid residues in a polypeptide chain.

The secondary structure describes the stable arrangements of amino acid residues that give rise to recurring structural patterns.

The tertiary structure refers to the 3-D folding of a polypeptide.

The quaternary structure describes a protein with two or more polypeptide subunits arranged in space.

1.2   Primary Structure

In the second half of the nineteenth century, it was found that proteins could be broken down by boiling them in water for a long time or treating them with acid or alkali. A few years later, Emil Fischer discovered that proteins consist of long chains of amino acid residues. As shown in Fig. 1.2, when two or more amino acids combine to form a peptide, water is removed during the formation of peptide bonds, leaving amino acid residues [1, 2]. This dehydration synthesis reaction (condensation reaction) commonly occurs between amino acids.

Thus, when many amino acids are joined, a polypeptide can be formed. Although the terms "protein" and "polypeptide" are sometimes used interchangeably, proteins typically have longer backbones with hundreds to several thousands of amino acid residues. Figure 1.3 shows the characteristic dimensions of a polypeptide chain derived from experimental results of amino acid and peptide structures [2, 3].

It appears that the peptide C-N bond is somewhat shorter (bond length 0.13 nm) than the C_α–N and C–C_α bonds. In fact, the C-N bond is approximately 10% shorter than that typically found in C–N amine bonds. Moreover, the N–C bond has an approximately 40% double-bond characteristic, indicating a resonance or a partial sharing of two pairs of electrons between the carbonyl oxygen and the amide nitrogen. Thus, a small electric dipole can be formed as follows: the oxygen exhibits a partial negative charge, and the hydrogen bonded to the nitrogen has a net

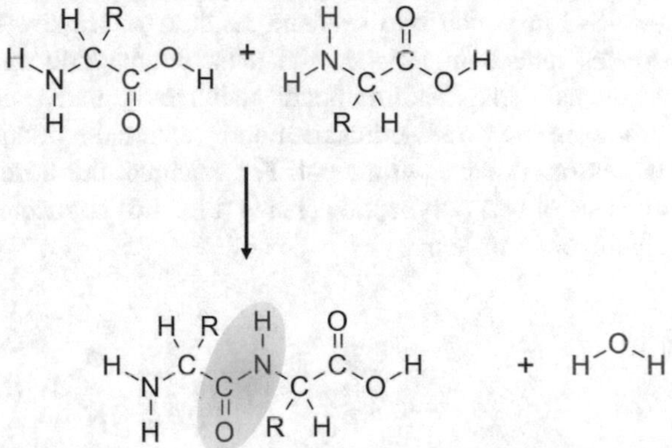

Fig. 1.2 Formation of a peptide bond by the condensation of two amino acids

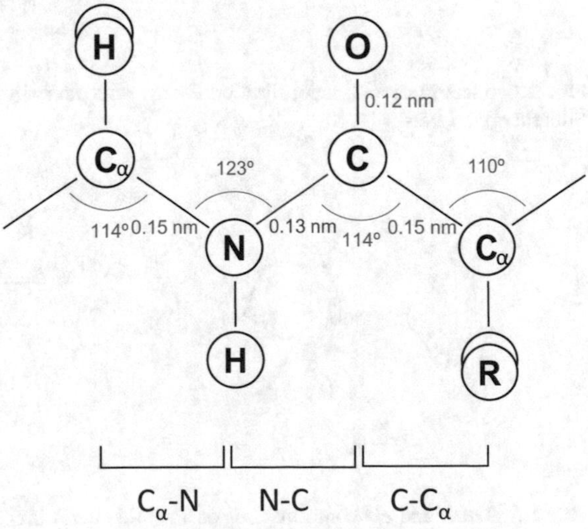

Fig. 1.3 Characteristic dimensions of a polypeptide chain. Adapted from Pauling [2, 3]

partial positive charge. Since each peptide bond has a double-bond characteristic due to this resonance, it cannot rotate freely. The peptide group is planar, and the range of possible conformations of a polypeptide chain is limited (Fig. 1.4).

In contrast, both the N–C$_\alpha$ and C$_\alpha$–C bonds can rotate, and therefore, the polypeptide chain can assume several configurations. Thus, a series of rigid planes share common points of rotation at C$_\alpha$. However, the configurations of proteins do not occur symmetrically; more than 99.95% of peptide bonds between amino acid residues are in the *trans* configuration since this configuration is favoured over the *cis* configuration. In contrast, regarding proline residues, the cyclic structure of the side chain indicates that both the *cis* and *trans* configurations have more closely equivalent energies. Thus, proline is found in the *cis* configuration much more frequently than other amino acids. Small proline-containing peptides in solution contain approximately 5–10% *cis* (syn) isomers (Fig. 1.5).

The primary structures of almost all intracellular proteins consist of linear polypeptide chains. However, many extracellular proteins contain *disulfide bonds* (–S–S–) in which two cysteine residues are linked by thiol groups which either creates intrachain links in the main polypeptide chain or links different chains together. These disulfide bonds stabilize the native conformations and protect the proteins from possible inactivation by external oxidants and proteolytic enzymes in the extracellular environment. For example, the hormone insulin is a protein that consists of two polypeptide chains (Fig. 1.6) connected by disulfide bridges, which stabilize the protein.

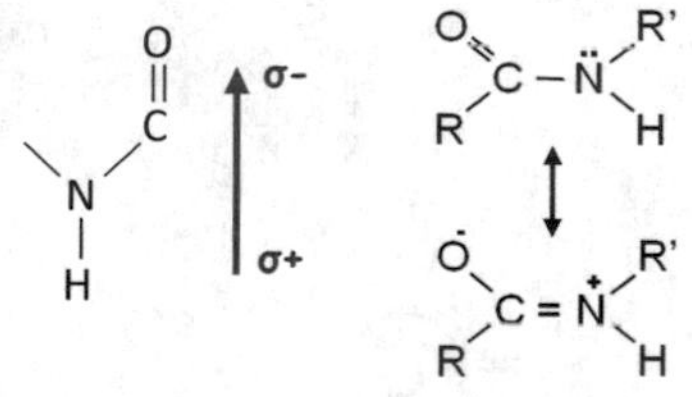

Fig. 1.4 Electric dipole, delocalization energy and planarity of peptide groups. Adapted from Silberberg and Fersht [4, 5]

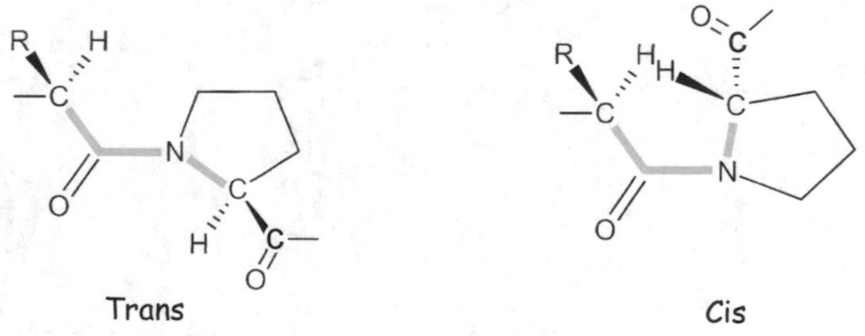

Fig. 1.5 *Trans* and *cis* conformations of a peptide bond involving the imino N of proline

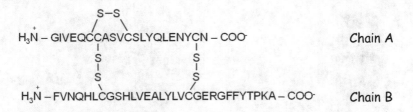

Fig. 1.6 Amino acid sequence of bovine insulin. The two polypeptides (chains A and B) are connected by disulfide cross-linkages (red)

Notably, to a large extent, the primary structure of proteins determines their spatial conformation. Determining the 3-D structure of a protein from genetic information is a complex task that scientists have investigated for years. In fact, DNA only contains information regarding the sequence of a protein's building blocks (namely, amino acid residues), which form chains. Predicting how these chains can fold into the 3-D structure of a protein is known as the "protein folding problem", which raises the question of *how a protein arrives at its native conformation in a short time.* For example, considering a polypeptide chain containing 100 amino acid residues that can adopt 10 different conformations on average, the polypeptide could theoretically have 10^{100} different conformations. If each conformation could be attempted in the shortest possible time ($\sim 10^{13}$ s, which is the time required for a singular molecular vibration), it would take approximately 10^{80} years to attempt all possible conformations! This problem was first described by Cyrus Levinthal in 1969 and is also known as the Levinthal paradox. If protein folding is a completely random process (trial-and-error approach), proteins would take longer than the age of the universe to obtain their native conformation. Such logic suggests that protein folding occurs via a hierarchical pathway in which small regions of secondary structures are assembled first and then gradually incorporated into larger structures.

Great computational efforts are currently ongoing to predict the stable 3-D structure of proteins starting from their amino acid sequences. These approaches include homology (or comparative) modelling, protein threading, and ab initio methods. In this scenario, recent developments in artificial intelligence and machine learning provide promising tools for the correct prediction of proteins.

1.3 Secondary Structure

Secondary structure refers to the 3-D conformation of local segments and is defined by the pattern of hydrogen bonds between the amino hydrogen and carboxyl oxygen atoms in the peptide backbone. Thus, protein configurations can be stabilized by N–H···O hydrogen bonds. The typical secondary structural elements of proteins are α-helices and β-sheets as follows:

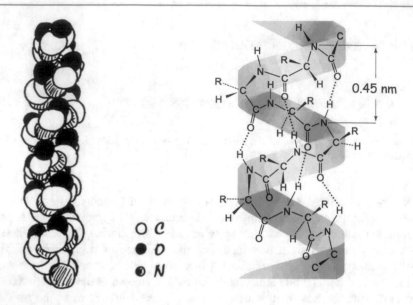

Fig. 1.7 Right-handed α-helix occurring in proteins (left). The circles represent the van der Waals radius of each atom. Adapted from Fersht [5]. *Courtesy of N. Lygeros.* The α-helix structure is stabilized by 3 chains of hydrogen bonds (right). Adapted from Georgiev and Glazebrook [6]. Reprinted with permission from Elsevier

(i) The α-helix, which is a common form of a protein secondary structure, is a right-hand-spiral conformation in which each N–H group forms a hydrogen bond with the C=O of the amino acid located three or four residues earlier along the protein sequence (N–H ⋯O) (Fig. 1.7). Left-handed α-helices are less energetically favoured and are not typically observed in proteins, representing another example of asymmetry in the structure of proteins occurring in nature.

Among the different types of local structures in proteins, the α-helix is the most regular, the most readily predicted from the sequence of amino acids, and the most common; approximately one-fourth of amino acid residues in proteins can be found in α-helices. The amino acids in an α-helix are arranged in a right-handed helical structure with a 100° turn in the helix and a translation of approximately 0.5 nm along the helical axis.

The electric dipole of a peptide bond is spread along the α-helix through intrachain hydrogen bonds, resulting in an overall helix dipole with carboxyl (σ) and amino (σ+) termini (Fig. 1.8). Then, an α-helix dipole arises from a set of interactions between local microdipoles, such as C=O⋯H–N [7, 8].

(ii) The β-conformation consists of β-strands connected laterally by at least two or three backbone hydrogen bonds, forming generally twisted, pleated sheets.

Fig. 1.8 Helix dipole (left) and electric dipole of a peptide bond (right). Adapted from Fersht [5]

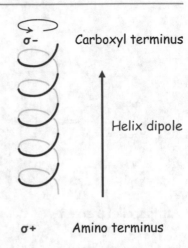

The zigzag structure of the polypeptide units leads to a pleated appearance of the overall sheet. Hydrogen bonds occur between adjacent units of polypeptide chains within the sheet. The adjacent polypeptide chains in a β-sheet can be either parallel or antiparallel, i.e., with the same or opposite amino to carboxyl orientations, respectively (Fig. 1.9).

In general, the backbone hydrogen bonds of β-sheets are considered slightly stronger than those found in α-helices and are less accessible by external water molecules.

Supersecondary structures (or motifs) consist of recurring substructures (or folds) due to possible combinations of α and/or β structures. An example is the β-hairpin, which is a simple protein structural motif involving two β-strands that appear similar to a hairpin. The motif consists of two strands that are adjacent and linked by a short loop of two to five amino acids (Fig. 1.10). This structure typically occurs in antiparallel β sheets.

The β-hairpin structure frequently appears in four β-strands connected in a motif called a Greek key because it is reminiscent of the Greek decorative motif or in six strands described as a jellyroll (Fig. 1.11).

The connections in parallel β-sheet motifs are more complex because the connectors have to change directions twice. The connections often consist of α-helices in β-α-β motifs as shown in Fig. 1.12.

In polypeptides, the energy of an N–H ··· O=C hydrogen bond is approximately 8 kcal mole^{-1}. Thus, the interaction between molecules and similar neighbouring molecules may cause small torques in the structures and slightly deform these structures into configurations with a different number of residues per turn. For instance, in a helix structure with 5.13 amino acid residues per turn, each amide group is hydrogen-bonded to the fifth amide group beyond it along the helix. Thus, the angle between the N–H and N–O vectors in the fifth amide hydrogen-bonded structure is approximately 25°.

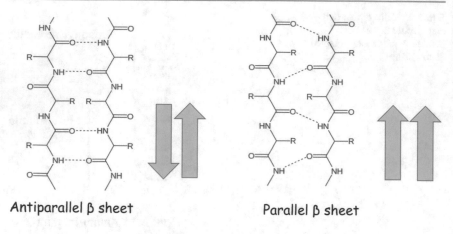

Antiparallel β sheet   **Parallel β sheet**

Fig. 1.9 Antiparallel and parallel β-sheet hydrogen bonding patterns

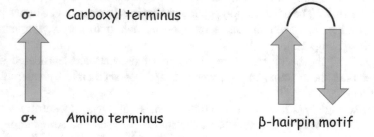

σ-   Carboxyl terminus

σ+   Amino terminus   **β-hairpin motif**

Fig. 1.10 β-hairpin motif

1.4 Tertiary Structure

Secondary structural elements typically form as intermediates before the protein folds into its tertiary structure. The protein tertiary structure consists of the 3-D arrangement of a protein, namely, a polypeptide chain "backbone" that interacts with one or more protein secondary structures (or protein domains). For example, Fig. 1.13 shows Top7, which is an artificial 93-residue protein designed by Brian Kuhlman and Gautam Dantas that is not found in nature. The interactions of side chains within a particular protein determine its tertiary structure. Thus, the α-helices and β-pleated sheets can be folded into a compact structure by hydrophobic interactions.

The tertiary structure is stable when the protein components are locked together by specific interactions, such as salt bridges, hydrogen bonds and disulfide bonds. The tertiary structure also strongly depends on the environmental conditions.

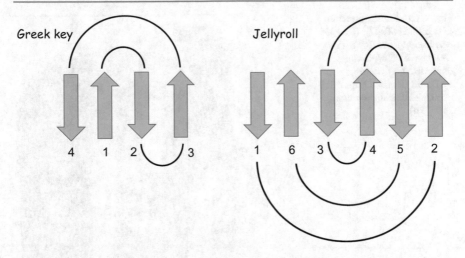

Fig. 1.11 Antiparallel β-sheet motifs arranged in the Greek key (left) and jellyroll (right) motifs. Adapted from Fersht [5]

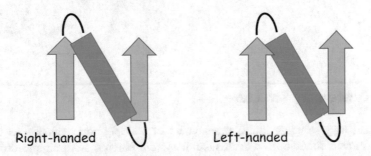

Fig. 1.12 β-α-β hairpin motifs. Adapted from Fersht [5]

For instance, the spherical (globe-like) structure of globular proteins is caused by the tertiary structure as these proteins consist of a core of hydrophobic amino acid residues and a surface of hydrophilic residues. The latter allows dipole–dipole interactions with water molecules, rendering globular proteins more soluble than fibrous or membrane proteins. Proteins may sequester potential reactive side chains in the interior of the globular structure, thereby slowing its degradation rates. However, peptides typically do not have an organized tertiary (or higher order) structure, and thus, side chains are fully solvent exposed (maximal chemical instability).

Fig. 1.13 Tertiary structure
of a protein (Top7). (*By kind
permission from RCSB PDB;
October 2005 Molecule
of the Month feature
by David Goodsell* https://doi.
org/10.2210/rcsb_pdb/mom_
2005_10)

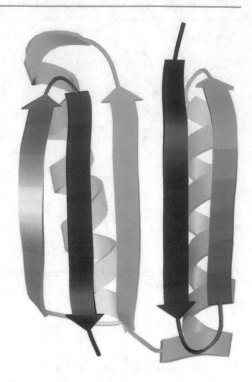

1.5   Quaternary Structure

The quaternary structure is the arrangement of multiple noncovalently associated folded protein subunits in a multisubunit system ranging from simple dimers to large oligomers and complexes with several subunits.

For example, a molecule of haemoglobin has four polypeptide subunits as follows: two identical α chains and two identical β chains. The quaternary structure of haemoglobin is the tetrahedral arrangement of its four subunits in a structure approximately 5.5 nm in size. These four polypeptide chains are held together via noncovalent bonds. Most of the 574 amino acids in haemoglobin form α-helices connected by short nonhelical segments. Hydrogen bonds stabilize the helical parts inside this protein, thereby causing attractions within the molecule that cause each polypeptide chain to fold into a specific shape as shown in Fig. 1.14.

In most vertebrates, haemoglobin is an assembly of four globular protein subunits. Each subunit consists of a protein chain tightly associated with a nonprotein prosthetic haem group. The latter comprises four pyrrole molecules cyclically linked together with an Fe ion in the centre (Fig. 1.15).

Iron is the site of oxygen binding and coordinates with four nitrogen atoms, which all lie in one plane. The Fe species can be in either the Fe^{2+} or the Fe^{3+} state, but ferrihaemoglobin (Fe^{3+}) cannot bind oxygen. Upon binding, oxygen

Fig. 1.14 Quaternary
structure of human
haemoglobin. The two α
subunits of haemoglobin are
shown in green, and the two β
subunits of haemoglobin are
shown in orange from PDB
code 1A3N [9]

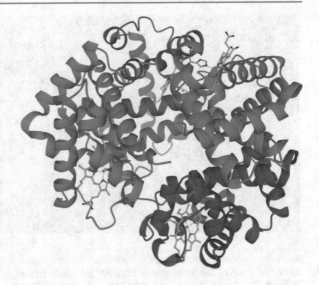

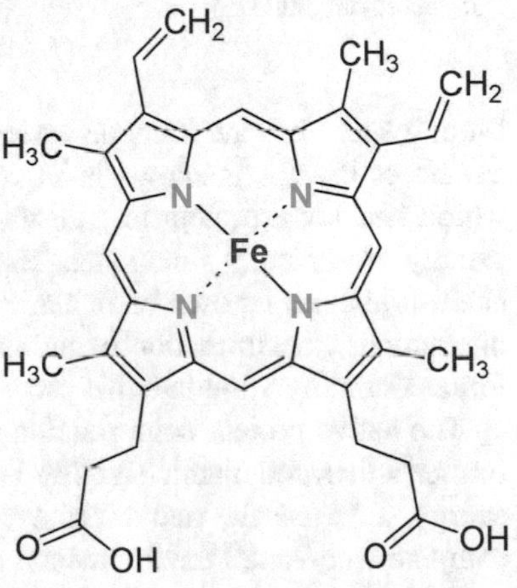

Fig. 1.15 Skeletal outline of
the haem B group

temporarily and reversibly oxidizes Fe^{2+} to Fe^{3+} while temporarily becoming superoxide ions. Figure 1.16 shows how the binding of oxygen changes the position of the iron ion. In the deoxy state (deoxyhaemoglobin), iron lies slightly outside the plane; in the oxy state (oxyhaemoglobin), iron lies within the plane.

1.6   Driving Forces in Protein Folding

The 3-D structure of a protein is critical for its activity. If a protein does not fold into the correct structure this usually results in inactive proteins, and misfolded proteins may have altered functionality. Many allergies are due to incorrect protein

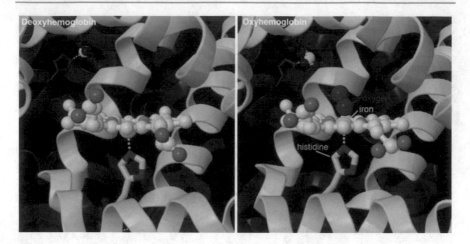

Fig. 1.16 Schematization of how the position of iron changes from slightly outside the plane to inside the plane of the haem group when oxygen binds haemoglobin. (*By kind permission from RCSB PDB; May 2003 Molecule of the Month feature by David Goodsell* https://doi.org/10.2210/rcsb_pdb/mom_2003_5)

folding such that the immune system fails to produce antibodies for some protein structures [5, 9]. As shown in Fig. 1.17, protein folding is a spontaneous process which enables a protein to gain its 3-D structure, namely, a conformation that is usually biologically functional. Therefore, protein folding is when a polypeptide chain folds into its own particular 3-D structure from a random coil. Hydrophobic interactions, the formation of intramolecular hydrogen bonds, and van der Waals forces normally stimulate this process.

The native protein conformation is more stable than the unfolded state, with the former's thermodynamic stability being only approximately 5–20 kcal/mol in free energy. Because the native protein structure shows low conformational stability, comparatively small environmental changes, such as temperature, pH or salt, in the protein-solvent may cause unfolding.

Cellular compartments contain specific proteins (chaperones) that may create interactions with the polypeptide chain to form the native 3-D structure. Chaperones stabilize a polypeptide in its folding pathway, thereby avoiding incorrect structural conformations. Since protein folding is thermodynamically favoured, this process has a negative Gibbs free energy. Overall, this process of folding may be considered a free-energy funnel (Fig. 1.18).

The funnel of a given protein may exhibit several shapes according to the number and variety of semistable intermediaries in these folding pathways. Any folding intermediary which exhibits significant stability and a limited lifespan may be considered a local free-energy minimum, a depression in the funnel. However, the unfolded state is distinguished by high levels of conformational entropy and comparatively high energy [4]. With the process of folding, the narrowing of the

Fig. 1.17 Protein structure before (random coil) and after the folding process. *Courtesy of N. Lygeros*

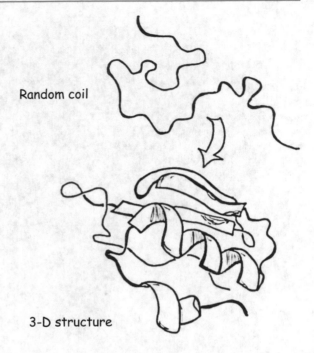

Random coil

3-D structure

funnel results in less conformational space to be sought as the protein nears its native state. At the funnel's lower section, a collection of unfolded intermediates becomes a single conformation (namely, the native conformation).

The folding funnel theory is associated with the hydrophobic collapse theory, which posits that the main force behind protein folding is the stabilization linked to the isolation of hydrophobic amino acid side chains inside the folded protein. Such stabilization enables the solvent to maximize its entropy, thus making the overall ΔG lower.

1.7 Structural Flexibility in Proteins

Many processes of a biochemical nature, for example signal transduction, protein transport, antigen recognition and enzyme catalysis, all depend on this ability to alter conformations or to adapt to change [12]. The dynamics within a protein enable its conformation to adapt and react to the environment. Proteins are all able to change, though they have varying extents of motion. The dynamism of proteins ranges from slight variations in side chains and movements of active site loops, to secondary structure fluctuations and re-arrangements of the whole protein fold [13]. As illustrated in Fig. 1.19, proteins exist as a collection of several arrangements (P_i), even if the predominant conformation is the native state (P_N) interacting with a ligand (L). Other conformers denote structural variations ranging from small side

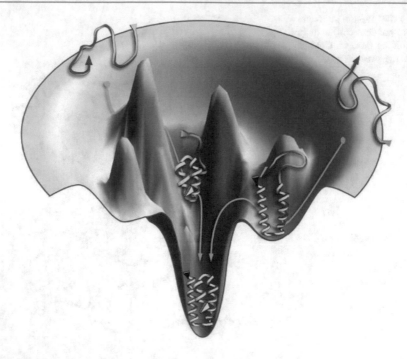

Fig. 1.18 The folding problem illustrated as a free-energy funnel landscape. A significant number of unfolded structures demonstrate high energy, but there are few low-energy folded structures. From Dill and MacCallum [11]. Reprinted with permission from AAAS

chain modifications and re-arrangements in the active site loop to significant fold changes. Thus, minor conformers (e.g., P_4) may promote other functions, such as indiscriminate interactions with L^* (another ligand that does not bind in the native conformation). Consequently, mutations can progressively alter these equilibria such that the scarcely populated conformer P_4 may become more favourable with considerable effects on the corresponding indiscriminate function. Therefore, functional modifications and adaptation to new environments may appear over time (protein evolvability).

Overall, a protein is flexible when it can adapt to different ligands or environments. The flexibility (or adaptability) of a protein may occur in small modifications (e.g., some amino acid side chains of a protein act to bind a ligand), or significant changes (e.g., the folding of some proteins is favoured by the existence of the proper ligand). Evidence of local flexibility can often be seen in NMR spectroscopy or electron density maps obtained by X-ray crystallography.

The crystallization process affects the protein structure, and dissimilar crystallization conditions may lead to various structural conformations. Figure 1.20 shows a collection of 44 crystal structures of hen egg white lysozyme, demonstrating that diverse conditions of crystallization bring about slightly different arrangements of a variety of surface-exposed loops and termini.

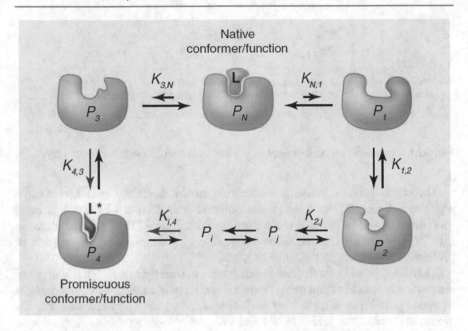

Fig. 1.19 Dynamics and evolvability in proteins. In the model proteins appear as a collection of conformations with the native state being dominant (e.g., P_N, interaction with the native ligand L). Other (minor) conformers (e.g., P_4) may perform other functions, such as indiscriminate interactions with L*, which is a ligand the protein did not evolve in order to bind. From Tokuriki and Tawfik [13]. Reprinted with permission from AAAS

Fig. 1.20 Collection of 44 crystal structures of hen egg white lysozyme. *Courtesy by N. Lygeros*

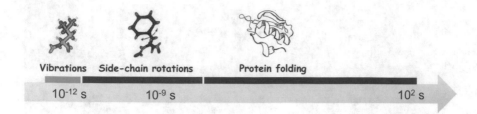

Vibrations   Side-chain rotations   Protein folding

10^{-12} s 10^{-9} s 10^2 s

Fig. 1.21 Conformational events occurring in proteins. Adapted from Teilum et al. [12]

The conformational events that constitute protein flexibility span 13 orders of magnitude (Fig. 1.21) as follows: the quickest events consist of covalent bond vibrations as well as fast side chain rotations on the pico- (10^{-12} s) to nanosecond (10^{-9} s) timescale, while the slowest events, e.g., protein–ligand dissociation or protein folding, occur in the range of hours.

Therefore, proteins are flexible structures due to their dynamics, but this does not consequently result in them being flexible. Flexibility is a fundamental requirement of native proteins to adapt to their environment. Some flexibility can appear in the amino acid sequence, which is not fixed or constant in a protein. For example, approximately 20–30% of proteins present in humans are polymorphic and display variations in their amino acid sequence with limited or no effect on their function [14].

As previously mentioned in Sect. 1.1, several human diseases are directly related to protein misfolding, which is a direct consequence of structural flexibility. Specifically, neurodegenerative diseases are linked to the creation of insoluble protein inclusions, often as protein fibrils. Conformational changes such as these require the unfolding of globular proteins or the partial folding of intrinsically disordered proteins. Many structurally disordered proteins interact with their binding partners via relatively short components with fluctuating complementary structures that become more ordered upon binding [15].

1.8   Thermodynamics of Protein–ligand Complexes

In order to study the thermodynamics of protein–ligand interactions and protein flexibility, consider the following protein (P) that binds a ligand (L):

$$P + L \leftrightharpoons PL$$

Under equilibrium conditions, the ratio between concentrations of the complex PL and free molecules P and L is provided by the equilibrium constant (K_a) as follows:

$$K_a = \frac{[PL]}{[P][L]}$$

K_a may be expressed as the difference in $\Delta G°$ between the complex PL and the free molecules P and L as follows:

$$\Delta G° = -RT\ln K_a$$

$\Delta G° < 0$ indicates that the formation of the complex PL is favoured, whereas the dissociation of PL is energetically more expensive.

According to Hess's Law, ΔG can be decomposed into changes in enthalpy (ΔH) and entropy (ΔS) as follows:

$$\Delta G = \Delta H - T\Delta S$$

The enthalpy change (ΔH) reflects the variation in the energy of the interactions that stabilize the protein structure.

When $\Delta H < 0$, new stabilizing interactions are created, and energy is discharged to the surrounding environment. Similarly, the change in entropy determines if the system will become more ordered ($\Delta S < 0$) or disordered ($\Delta S > 0$). When a non-covalent interchange involving a protein and a ligand is created, the enthalpy of the complex decreases ($\Delta H < 0$). Surrounding the binding interface, the system loses conformational freedom, leading to a decrease in the entropy of the molecules ($\Delta S < 0$). As more interactions form, the complex becomes more ordered. In fact, a direct correlation between ΔH and ΔS has been reported in many protein–ligand interactions [12].

Changes in entropy during protein binding processes and protein conformational modifications are complex phenomena. Overall, the binding of a ligand to a protein might produce an increase or a decrease in entropy. In fact, ΔS is a combination of translational and rotational entropy ($\Delta S_{transl-rot}$), conformational entropy (ΔS_{conf}) and solvent entropy (ΔS_{solv}) as follows:

$$\Delta S = \Delta S_{transl-rot} + \Delta S_{conf} + \Delta S_{solv}$$

- $\Delta S_{transl\ rot}$ depends on the nature of the ligand and the stabilizing interactions.
- ΔS_{conf} depends on the conformational space available to the protein.
- ΔS_{solv} results from the change in the solvent-exposed area upon binding and is directly linked to the hydrophobic effect. The hydration of a nonpolar surface has high entropy, while polar surfaces typically exhibit negligible ΔS_{solv}. In fact, the water molecules filling the active site of a protein and surrounding the ligand are important for determining the binding thermodynamics. Thus, both proteins and ligands are stabilized by hydrophobic effects that favour the aggregation of nonpolar regions.

ΔS_{conf} and ΔS_{solv} dominate ΔS with opposing contributions of the same order of magnitude, whereas $\Delta S_{transl\text{-}rot}$ typically has a negligible effect. Nevertheless, which ΔS predominates depends on the molecular structures involved in the system.

1.9   Characterization of Protein Structures [16]

The architecture of proteins is an object of study in structural biology. Over the last few decades, studies in this branch of molecular biology developed complementary characterization techniques for the determination of the 3-D structure of proteins, such as X-ray crystallography, NMR spectroscopy and cryo-electron microscopy (cryo-EM). Proteins can be physically unstable, and aggregates may occur during protein production, purification and storage. Therefore, several spectroscopic techniques have been applied to investigate protein aggregation, including FTIR spectroscopy (effective in detecting β-sheet structures), Raman spectroscopy (used to predict aggregation and particle formation), NMR (used to monitor protein aggregation at high concentrations) and extrinsic fluorescence spectroscopy (sensitive to the formation of ordered aggregates).

1.9.1   X-Ray Crystallography

Most of the structures deposited in the Protein Data Bank (PDB database) were determined by X-ray crystallography. Indeed, the determination of structure from X-ray diffraction patterns provides information regarding the electron density distribution within a molecule [17]. James Batcheller Sumner was a pioneer who applied this technique to enzymology and, in 1926, demonstrated that enzymes (in this case urease) can be isolated, purified and then crystallized. The crystal is subjected to an intense beam of X-rays, and the diffraction of the X-rays generates diffraction patterns that, in principle, allow us to determine the precise electron density distribution in any protein. The electronic density map that is obtained can be elaborated with computational models to determine the corresponding atomic locations. In the 1930s, this characterization technique was used to map increasingly large and complex molecules. In 1948, during an important symposium on haemoglobin organized in Cambridge, M. Perutz and J.C. Kendrew revealed the possible application of X-ray crystallography to the study of haemoglobin and myoglobin [18].

Currently, the degree of resolution achievable with the crystallographic approach can be very high (< 0.15 nm). X-ray analyses can provide detailed information regarding the position of each atom in a protein with atomic details of the ligands, inhibitors, ions or other components incorporated in the crystal. This technique represents an excellent method for determining rigid protein structures that form ordered crystals but is not suitable for flexible proteins, which are invisible in electronic crystallographic density maps.

1.9.2 Nuclear Magnetic Resonance Spectroscopy

NMR spectroscopy is currently used to determine the structure of purified proteins kept in solution under conditions closer to physiological conditions. This technique allows resolutions < 0.2 nm. Thus, NMR is the method of choice for studying the atomic structure and internal dynamics of flexible proteins, conformational equilibria and protein–protein interactions. Using this technique, the purified protein is immersed in a strong magnetic field and subjected to radio waves. NMR results can be analysed to obtain a list of atomic nuclei that are close to each other and characterize the local distribution of atoms. These data can be used to construct a model of the protein. The application of this characterization technique is currently limited to proteins of a small to medium size (< 40 kDa) since large proteins exhibit problems of overlapping peaks in NMR spectra.

1.9.3 Cryo-Electron Microscopy

The Nobel Prize in Chemistry 2017, which was awarded to the scientists Jacques Dubochet, Joachim Frank and Richard Henderson, recognizes the recent evolution of cryo-EM for determining high-resolution structures of biomolecules in solution. Three-dimensional structures of a molecule can be obtained from micrographs (2-D projections) by the combination of transmission electron microscopy (TEM) and computerized data processing. A currently promising technique consists of the acquisition of many images of a protein in its native state, which is frozen directly in the extraction buffer, thus resulting in a thin layer of noncrystalline ice. The images provide several views of the biomolecule in different orientations, which then, via a "single-particle analysis", allows us to obtain a 3-D map of its electronic density. Thus, due to the development of calculation algorithms and new detectors with high sensitivity, the combination of cryo-EM and a single-particle analysis provides 3-D structures of proteins with resolutions comparable to crystallographic ones, thereby allowing the visualization of amino acid side chains and the localization of components.

1.10 Summary

- Proteins can be structurally described at different levels of complexity.
- The primary structure determines the 3-D conformation of a protein. Predicting the 3-D arrangement of a protein from its primary structure is a complex task known as the protein folding problem.
- The folding process can be thermodynamically described as a free-energy funnel. The native structure of a protein lies at the lowest free-energy point (bottom of the funnel), whereas the unfolded states are characterized by a high level of conformational entropy and relatively high energy.

- Proteins are dynamic and flexible structures. The timescale of the conformational events that constitute protein flexibility spans several orders of magnitude (from bond vibrations and side chain rotations to protein folding).

1.11   Questions

1. What is a common feature of both an α-helix and a β-pleated sheet?
2. What is a common property of the tertiary structure and quaternary structure of a protein?
3. What are the main noncovalent interactions thought to be involved in stabilizing protein folding and protein–ligand binding interactions in solution? What is a possible effect of water on these interactions?
4. Flexibility is a fundamental requirement for proteins to adapt to their environment. However, several human diseases are related to protein misfolding, which is a direct consequence of structural flexibility. In which cases does flexibility promote protein activity? In contrast, when does protein activity worsen due to structural flexibility?

References

1. L. Pauling, R.B. Corey, R. Hayward, The structure of protein molecules. Sci. Am. **191**(1), 51–59 (1954)
2. L. Pauling, *The Nature of Chemical Bond*, 3rd edn. (Cornell University Press, New York, 1960)
3. L. Pauling, R.B Corey, H.R. Branson, Proc. N.A.S. **37**, 205–211 (1951)
4. M.S. Silberberg, *Chemistry*, 5th edn. (McGraw-Hill, New York, 2009)
5. A. Fersht, *Structure and Mechanism in Protein Science* (W. H. Freeman, New York, 1998)
6. D.D. Georgiev, J.F. Glazebrook, Physica A **517**, 257–269 (2019)
7. J.J. He, F.A. Quiocho, Protein Sci. **2**(10), 1643–1647 (1993)
8. E.J. Milner-White, Protein Sci. **6**(11), 2477–2482 (1997)
9. B. Alberts, D. Bray, K. Hopkin, A. Johnson, J. Lewis, M. Raff, K. Roberts, P. Walter *"Protein Structure and Function". Essential Cell Biology*, 3rd edn. (Garland Science, New York, 2010)
10. J.R. Tame, B. Vallone, Acta Crystallogr. Sect. D **56**, 805–811 (2000)
11. K.A. Dill, J.L. MacCallum, Science **338**, 1042 (2012)
12. K. Teilum, J.G. Olsen, B.B. Kragelund, Cell. Mol. Life Sci **66**, 2231–2247 (2009)
13. N. Tokuriki, D.S. Tawfik, Science **324**, 203–207 (2009)
14. D. Nelson, M.M. Cox, *Lehninger Principles of Biochemistry*, 7th edn. (Macmillan, 2017), pp. 75–105
15. T. Mittag, L.W. Kay, J.D. Forman-Kay, J. Mol. Recognit. **23**, 105–116 (2010)

16. M. Piumetti, C. Pagliano, Gli enzimi e l'industria, Aracne 21–29 (2018)
17. J.M. Thomas, *Architects of Structural Biology* (Oxford University Press, Oxford, 2019)
18. M.F. Perutz, J.C. Kendrew, in *Haemoglobin* ed. by F.J.W. Roughton, J.C. Kendrew (Butterworths, London, 1949), p. 161

Enzymes and Their Function

2

Marco Piumetti and Andrés Illanes

2.1 Introduction

Similar to all living organisms, human life depends on the action of biological catalysts called enzymes. Each living process requires (evolved) enzymes. Enzymes are globular proteins with catalytic activity responsible for the chemical reactions of cell metabolism. The effects of enzymes have been recognized since ancient times. The fermentation of sugar to alcohol by yeast is an early example of a biotechnological process that involves enzymes. However, the physico-chemical properties of enzymes have only recently been investigated properly with the combination of concepts from protein chemistry, molecular biophysics, and molecular biology.

Currently, more than 5000 biochemical reactions are known to be catalysed by enzymes. The molecules upon which enzymes act (reactants) are called substrates, which are converted into molecules known as products.

Each reaction of cell metabolism requires an enzyme to catalyse it at a rate fast enough to sustain life. Enzymes are highly efficient catalysts; enzymes increase the reaction rates by 10^8 to 10^{20} times and are extremely specific (each reaction is generally catalysed by a specific enzyme). This remarkable specificity results from the amino acid residues that constitute the active site. For example, the urease enzyme of *Helicobacter* (Fig. 2.1) catalyses only the hydrolysis of urea in water to produce ammonia and carbonic acid:

$$(NH_2)_2C = O + 2H_2O + H^+ \rightarrow 2NH_4^+ + HCO_3^-$$

Electronic Supplementary Material The online version of this chapter (https://doi.org/10.1007/978-3-030-88500-7_2) contains supplementary material, which is available to authorized users.

Fig. 2.1 Urease enzyme in *Helicobacter pylori* from PDB code 1E9Z [1]

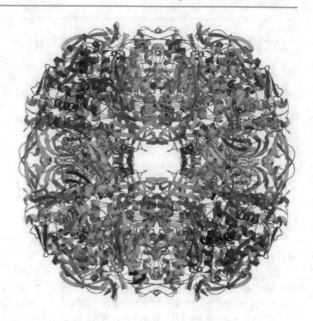

The rate constant of the uncatalysed reaction at room temperature is approximately $3 \times 10^{-10} \text{ s}^{-1}$, whereas in the presence of urease, the rate constant is $3 \times 10^4 \text{ s}^{-1}$, which is 10^{14} times greater than the rate of the uncatalysed reaction. None of the thousands of other enzymes present in the cell are active for this reaction.

It has been estimated that there may be 20,000–30,000 enzymes in the human body, and each enzyme is constructed in such a way that it can act as an effective catalyst for a particular chemical reaction that is inherent to the organism. Enzymes have undergone millions of years of natural selection and continue to perform their roles as biological catalysts with great efficiency. Therefore, understanding the structural properties and functions of enzymes is extremely useful for the design of effective catalysts for environmental and energy-related applications.

2.2 Structure of Enzymes

Most biological catalysts are proteins (enzymes), although a few catalysts are ribozymes (ribonucleic acid catalysts), i.e., RNA molecules that can catalyse specific biochemical reactions. Ribozymes perform a much more limited set of reactions than enzymes. The native enzyme conformation is essential for catalytic activity. When an enzyme is denatured by dissociation into its units or even further by breaking down into its component peptides and amino acids, its catalytic activity is lost. Therefore, the 3-D conformation of an enzyme is essential for its catalytic activity.

Similar to other proteins, enzymes typically exhibit complex structures with molar masses ranging from 12,000 to more than 1,000,000 g/mol [2]. These biocatalysts are highly substrate specific, and the conversion of a substrate into a product is assumed to occur at a particular site on the enzyme molecule known as the active site.

Some enzymes do not require additional chemical groups for activity other than their amino acid residues. However, other enzymes need an additional chemical component called a cofactor, which is either one or more inorganic ions (e.g., Cu^{2+}, Fe^{2+} or Fe^{3+}, K^+, Mg^{2+}, Mn^{2+}, Ni^{2+} and Zn^{2+}) or a complex organic (e.g., nicotinamide adenine dinucleotide, NAD) or metallorganic (coenzyme) compound. A cofactor or metal ion that is tightly and even covalently bound to an enzyme is called a prosthetic group. Some enzymes may require both a coenzyme and one or more metal ions for their activity. As a result, a catalytically active enzyme and its bound coenzyme and/or metal ion are called a holoenzyme, whereas the protein part is called an apoenzyme (Fig. 2.2). A cofactor tightly binds an enzyme and is not altered during catalysis, and its behaviour is catalytic, while a coenzyme loosely binds an enzyme and is modified during catalysis, and its behaviour is stoichiometric.

Protein X-ray crystallography has revealed that a given enzyme has a unique and well-defined 3-D structure. Generally, hydrophobic amino acid residues tend to be associated with the hydrophobic interior of a folded molecule, whereas charged amino acid residues are usually located on the hydrophilic exterior of an enzyme. In the 1960s, preliminary high-resolution crystallographic studies of egg-white lysozyme confirmed these assumptions [3]. In fact, the active site of egg-white lysozyme is located in a cleft of the structure, which has subsequently proven to be a common feature of active sites in enzymes.

Fig. 2.2 Components of a holoenzyme

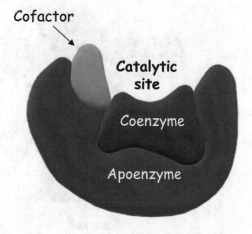

2.3 Enzymatic Process

Only a small portion of the enzyme structure is directly involved in catalysis, which occurs next to one or more substrate binding sites where the amino acid residues orient the substrate. The catalytic site and binding site comprise an enzyme's active site. In most cases, substrates bind the active site through intermolecular forces, such as H-bonds, dipole forces and other weak interactions.

Substrate binding

Enzymes must bind their substrates before catalysing any chemical reaction; thus, enzymes are usually stereospecific to their substrates. Specificity is achieved by binding pockets with shape, charge and hydrophilic/hydrophobic characteristics complementary to those of the substrate. Therefore, complementarity is necessary for affinity between an enzyme and a substrate (Fig. 2.3). Complementarity must be considered in both electric and geometric terms. Moreover, the active site exhibits a preorganized electrostatic environment that stabilizes the charge distribution of the substrate in the transition state.

To explain the specificity of enzymes, in 1894 Emil Fischer proposed a model in which an enzyme and a substrate have complementary shapes that fit precisely into one another. This is often referred to as the lock-and-key model [4, 5], where the lock is the enzyme, and the key is the substrate. Only a correctly sized and shaped key (substrate) fits the keyhole (active site) of a lock (enzyme). This model explains enzyme specificity but fails to explain the stabilization of the transition state that an enzyme achieves.

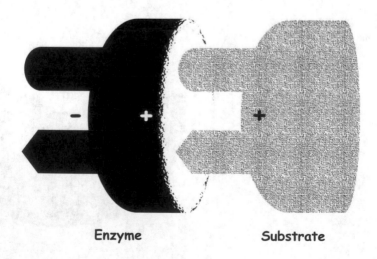

Enzyme   **Substrate**

Fig. 2.3 Complementarity between an enzyme and a substrate

Subsequently, in 1958, Daniel Koshland proposed a variation to the lock-and-key model [6]. In fact, since enzymes are flexible structures, the active site is continuously reshaped by interaction with the substrate. A scheme of the induced-fit model is shown in Fig. 2.4. The latter model incorporated Fisher's concepts of the complementarity of an enzyme and a substrate but introduced the concept of a flexible enzyme. The substrate-enzyme interaction appears as "the fit of a hand in a glove" [6].

Enzyme catalytic cycle

The enzyme catalytic cycle is a multistep reaction mechanism.

Since enzymes can be regenerated, catalytic cycles are usually described as a sequence of elementary steps that form a loop. In such loop, the initial step entails the binding of one or more substrates to the enzyme, whereas the final step consists of the release of the product(s) and enzyme regeneration. Thus, a single cycle includes all events that occur to an enzyme beginning in a specific state until it returns to that state after catalysing a reaction as follows:

$$\text{Enzyme}(E) + \text{substrates } (S_i) \rightarrow \text{enzyme}(E) + \text{product}(s)(P_i)$$

The main steps of an enzyme catalytic cycle are as follows:

1. The substrate binds the enzyme in a favourable manner.
2. This binding induces the enzyme to alter its shape to more closely fit the substrate.
3. When the active site of the enzyme is in proximity to the substrate, the chemical bonds of the substrate are broken, and a new enzyme-product complex is formed. This process describes a degradation reaction, e.g., hydrolysis, but does

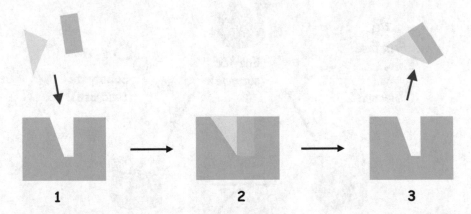

Fig. 2.4 Schematic illustration of enzyme flexibility in action according to the induced-fit mechanism. 1. Two substrates reach the cleft of the enzyme; 2. The enzyme changes its shape, forcing the substrate molecules to combine; 3. The product is released and the cleft of the enzyme returns to its initial shape

not represent the events occurring in a synthesis reaction, where an enzyme allows the formation of a bond between two molecules.

4. The enzyme releases the products, is regenerated and is ready to interact with another substrate molecule.

Figure 2.5 shows the case of invertase, which catalyses the hydrolysis of sucrose into fructose and glucose. In this case, the following events occur:

1. The enzyme is free to bind the substrate (sucrose).
2. Sucrose binds the enzyme, thereby forming an enzyme-sucrose complex.
3. The binding of sucrose to invertase destabilizes the glucose-fructose bond, and the bond breaks.
4. Products are released, and the enzyme is restored for a new cycle.

Enzymes perform catalysis by reducing the activation energy using different strategies [7–11]. For example, trypsin proteases perform covalent catalysis by using a catalytic triad (a group of three coordinated amino acids, present in their active site), stabilize the build-up of charge on the transition states by using an oxyanion hole, and complete hydrolysis by using an oriented water substrate.

Enzyme dynamics

Since many enzymatic reactions occur on time scales ranging from microseconds to milliseconds, the conformational dynamics of enzymes at these time scales might be linked to their catalytic action [12]. As previously reported in proteins, enzymes

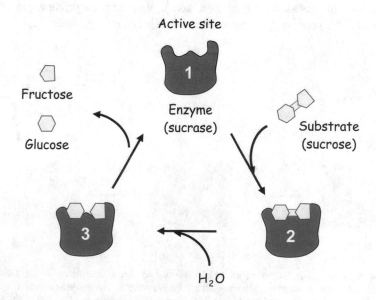

Fig. 2.5 Catalytic cycle of the hydrolysis of sucrose by invertase

are not static structures; in fact, enzymes are highly flexible molecules and exhibit dynamic motions. Therefore, movements of the enzyme structure involving individual amino acid residues, groups of residues forming a protein loop or unit of a secondary structure, or even an entire protein domain are possible.

Thermodynamics of enzyme-catalysed reactions

Enzymes increase the rates of spontaneous S $\rightarrow$ P reactions by decreasing the energy of the transition state. Firstly, a low-energy enzyme–substrate complex [ES] is formed. Secondly, the enzyme stabilizes the transition state [ES‡] so that it needs less energy than the uncatalysed reaction. In the final step, the enzyme-product complex [EP] dissociates to release the products (Fig. 2.6).

Enzymes can link two or more reactions as follows: A thermodynamically unfavourable reaction can be driven by a thermodynamically favourable reaction, such that the combined energy of the products is less than that of the substrates. For example, the hydrolysis of adenosine triphosphate (ATP) may promote other biochemical reactions. ATP hydrolysis is the catabolic process whereby the chemical energy is released by splitting the high-energy phosphoanhydride bonds. The reaction products are adenosine diphosphate (ADP) and orthophosphate (P_i). ADP may be hydrolysed further to yield energy, adenosine monophosphate (AMP), and another P_i. The hydrolysis of the phosphate groups in ATP is markedly exergonic ($\Delta G < 0$) because multiple resonance structures greatly stabilize the resulting P_i group, rendering the energy of ADP (and P_i) far lower than that of ATP. Figure 2.7 shows the structures of ATP and ADP and the resonance structures of P_i.

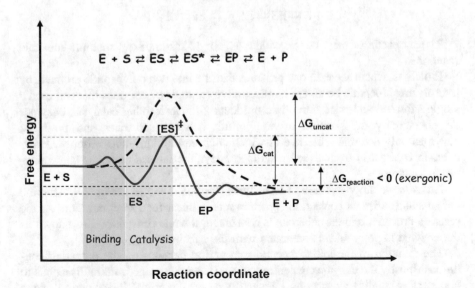

Fig. 2.6 Reaction coordinate diagram for an enzyme-catalysed reaction (read line) versus the uncatalysed reaction (dashed line)

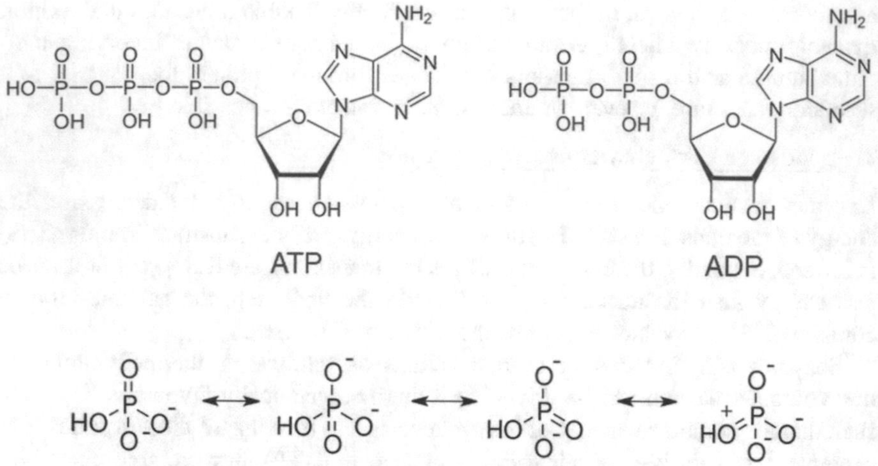

Fig. 2.7 ATP and ADP structures and four possible resonance structures of orthophosphate (P_i)

2.4 Enzyme Kinetics

Typically, enzymes can convert substrates into products via a series of steps that constitute the enzymatic reaction mechanism as follows:

$$E + S \rightleftarrows ES \rightleftarrows ES * \rightleftarrows EP \rightleftarrows E + P$$

These reactions may be classified as single-substrate or multiple-substrate reactions.

Proteases, which separate one protein substrate into two polypeptide products by hydrolysing a peptide bond, are an illustrative example of enzymes that can bind a single substrate and release multiple products. Similar to other catalysts, enzymes do not modify the position of equilibrium between substrates and products; enzymes only determine the rate at which such equilibrium is approached. However, in contrast to uncatalysed reactions, enzyme-catalysed reactions show saturation kinetics.

An example progress curve of an enzyme reaction is shown in Fig. 2.8. After the enzyme and substrate contact, the reaction rate is linear for a short duration. As the reaction proceeds and the substrate is consumed, the rate constantly slows down as the enzyme is progressively saturated with the substrate.

The duration of the initial reaction rate period depends on the operating conditions (mostly the enzyme concentration used) and can range from milliseconds to hours. Most studies on enzyme kinetics focus on this initial linear period of the reaction. Nevertheless, it is possible to follow the whole reaction time curve and fit these data with a nonlinear rate equation by a progress-curve analysis.

Fig. 2.8 Product formation over time during an enzyme reaction

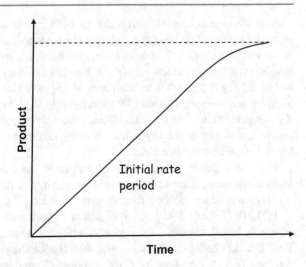

2.4.1   Single-Substrate Reactions

Enzymes with single-substrate mechanisms include isomerases (e.g., triosephosphate isomerase and bisphosphoglycerate mutase), intramolecular lyases (e.g., adenylate cyclase) and ribozymes (e.g., RNA lyase) [13]. Hydrolytic reactions involve two substrates but can be considered one-substrate reactions since the second substrate (water) which is also the solvent, is present in great excess (close to 55 M) and, therefore, does not affect the kinetics, which depend only on the other substrate (the solute).

However, some enzymes interact with a single substrate but perform different mechanisms (vide infra).

2.4.1.1   Michaelis–Menten Kinetics

The Michaelis–Menten equation is a well-known model of enzyme kinetics. This equation relates the initial reaction rate (v), namely, the rate of the formation of product P (or the rate of substrate consumption), to [S] the concentration of the substrate (S). Thus, the Michaelis–Menten equation is as follows:

$$v = \frac{d[P]}{dt} = -\frac{d[S]}{dt} = \frac{V_{max}[S]}{K_M + [S]}$$

Here, V_{max} represents the maximum reaction rate, which is attained at the saturating substrate concentration. As [S] increases, the enzyme molecules become saturated with the substrate, and the initial reaction rate approaches V_{max}. Actually, according to the Michaelis–Menten equation, V_{max} is a limit theoretically attained as [S] approaches infinity; in practice, [S] is limited by the solubility of S in the reaction medium; thus, the maximum reaction rate that can be obtained experimentally can be significantly lower than V_{max}.

The Michaelis constant (K_M) is the dissociation constant of the enzyme–substrate into free enzyme and substrate, which represents an inverse measure of the affinity of the enzyme for its substrate as follows: a small K_M indicates high affinity, and the reaction rate will approach V_{max}/K_M at low [S]. According to the Michaelis–Menten equation, K_M corresponds to the value of [S] at which the reaction rate is half of V_{max} as can be easily deduced by letting $v = \frac{1}{2} V_{max}$ in the equation. The values of K_M range from 10^{-2} to 10^{-5} M. These values depend on both the enzyme and the substrate and vary according to the operating conditions, where temperature and pH are the most relevant variables.

Enzymatic reactions with a single substrate are assumed to follow Michaelis–Menten kinetics. Therefore, these reactions exhibit a reaction rate with a saturation curve as a function of the substrate concentration (Fig. 2.9).

In 1913, Leonor Michaelis and Maud Leonora Menten proposed a model of invertase, which catalyses the hydrolysis of sucrose into glucose and fructose [14]. This kinetic model involves an enzyme (E) binding a substrate (S) to form a complex (ES), which then leads to a product (P) and regenerates the enzyme. This cycle can be represented schematically as follows:

$$E + S \underset{k_2}{\overset{k_1}{\rightleftarrows}} ES \overset{k_3}{\rightarrow} E + P$$

where k_1, k_2, and k_3 correspond to the forward, reverse and catalytic rate constants, respectively; the double arrow between S (substrate) and ES (enzyme–substrate complex) represents the reversibility of enzyme–substrate binding, and the single forward arrow represents the irreversible formation of the product from ES. The reaction from ES to E + P was considered irreversible because early experiments referred to hydrolysis reactions in aqueous media where, because of the mass action law, the reaction was completely displaced in the forward direction; this case does not apply to other types of reactions, such as isomerization.

Fig. 2.9 Michaelis–Menten saturation curve of an enzyme reaction showing the relationship between the substrate concentration [S] and initial reaction rate (v)

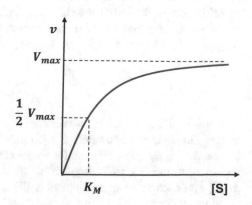

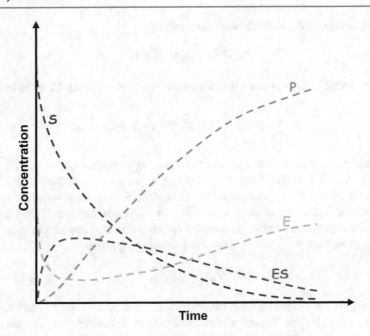

Fig. 2.10 Concentrations of enzyme E, substrate S, complex ES and product P over time

Figure 2.10 shows the evolution of the concentrations of enzyme E, substrate S, complex ES and product P over time in a closed system (batch reactor) assuming the presence of a kinetically controlling step (ES → E + P).

These trends can be described by a system of nonlinear ordinary differential equations that define the rate of change in these components over time as follows:

$$\frac{d[E]}{dt} = -k_1[E][S] + k_2[ES] + k_3[ES]$$

$$\frac{d[S]}{dt} = -k_1[E][S] + k_2[ES]$$

$$\frac{d[ES]}{dt} = k_1[E][S] - k_2[ES] - k_3[ES]$$

$$\frac{d[P]}{dt} = k_3[ES]$$

Although the enzymatic mechanism of ES → E + P can be quite complex, there is typically one rate-determining enzymatic step that allows this reaction to be modelled as a single catalytic step with an apparent unimolecular rate constant k_3. Moreover, if the reaction proceeds in the presence of one or more intermediates, k_3 appears as a function of the elementary rate constants.

In addition to the bi-steady state assumption,

$$E + S \rightleftharpoons ES \rightleftharpoons E + P,$$

it is also possible to consider more intermediates (e.g., P binds E) as follows:

$$E + S \underset{k_2}{\overset{k_1}{\rightleftharpoons}} ES \underset{k_4}{\overset{k_3}{\rightleftharpoons}} EP \underset{k_6}{\overset{k_5}{\rightleftharpoons}} E + P$$

In this case, further kinetic constants must be considered. In contrast, in the simplest case of a single elementary reaction (e.g., no intermediates), the kinetic constant is identical to the elementary unimolecular rate constant. The apparent unimolecular rate constant k_3 is also called the turnover number (TON) and represents the maximum number of catalytic cycles performed per unit of time. Thus, the maximum rate V_{max} results from the following equation:

$$V_{max} = k_3[E]_0$$

where $[E]_0$ is the initial enzyme concentration. This rate is approached when all enzyme molecules are bound to the substrate. In summary, the following two different regimes can be observed during an enzymatic reaction:

(a) When the substrate concentration is low ($[S] \ll K_M$), the Michaelis–Menten equation becomes

$$v = \frac{V_{max}[S]}{K_M} = \frac{k_3[E]_0[S]}{K_M}$$

Under these conditions, the reaction rate varies linearly with the substrate concentration (first-order kinetics).

(b) When the substrate concentration is high ($[S] \gg K_M$), the reaction rate appears to be independent of $[S]$ and asymptotically approaches the value of V_{max}.

$$\lim_{[S] \to \infty} v = V_{max} = k_3[E]_0$$

Additional amounts of substrate do not increase the reaction rate (zero-order kinetics). A scheme of these two regimes is shown in Fig. 2.11.

The Michaelis–Menten model is based on the assumption that an enzyme and substrate form a complex (ES) in a reversible process. The substrate bound to the enzyme in the ES complex is then converted into the product at the reaction rate determined by k_3.

Substrate

Enzyme

Low [S]   High [S]

Fig. 2.11 Enzymatic reactions under different regimes, namely, low and high substrate concentrations [S]

Under the conditions commonly used for enzyme activity measurement, the complex concentration [ES] can be considered constant during the observed reaction period (steady-state assumption as proposed by Briggs and Haldane) as follows [15]:

$$R_{ES} = k_1[S][E] - k_2[ES] - k_3[ES] = 0$$

where R_{ES} represents the complex formation rate under steady-state conditions.

By considering the total concentration of the enzyme

$$[E]_{tot} = [E] + [ES]$$

and the above two equations,

$$k_1[S][E]_{tot} - k_1[S][ES] - k_2[ES] - k_3[ES] = 0$$

such that

$$[ES] = \frac{k_1[S][E]_{tot}}{k_1[S] + k_2 + k_3}$$

By introducing the Michaelis constant K_M,

$$K_M = \frac{k_2 + k_3}{k_1}$$

The reaction rate as a function of the enzyme–substrate complex concentration is as follows:

$$v = k_3[ES]$$

The maximum reaction rate (V_{max}) corresponds to the reaction rate when all of the enzyme is saturated with the substrate, namely, $[ES] = [E]_{tot,}$ as follows:

$$V_{max} = k_3[E]_{tot}$$

Thus, the so-called Michaelis–Menten equation can be obtained as follows:

$$v = k_3[ES] = k_3 \frac{k_1[S][E]_{tot}}{k_1[S] + k_2 + k_3} = \frac{V_{max}[S]}{K_M + [S]}$$

This final equation shows the dependency of the reaction rate on the substrate concentration. It is worthwhile to compare this equation with the previous formulation of the M-M equation. The difference is in the meaning of K_M. According to the original rapid equilibrium hypothesis of M-M, K_M is simply an equilibrium constant (k_2/k_1), while in the Briggs-Haldane hypothesis, K_M is a dissociation constant, namely, $(k_2 + k_3)/k_1$, that obviously reduces to k_2/k_1 in rapid equilibrium, that is, when $k_2 \gg k_3$.

The plot of the reaction rate (v) versus [S] according to the Michaelis Menten equation is not linear but is actually a hyperbola. This nonlinearity could make it difficult to estimate the values of K_M and V_{max} accurately. Therefore, several linearization forms of the Michaelis–Menten equation have been proposed, such as the Lineweaver–Burk plot or double-reciprocal plot, which is a common way used to represent kinetic data in terms of the inverse of v versus the inverse of [S]. This is a linear form of the M-M equation and produces a straight line with a y-intercept corresponding to $1/V_{max}$ and an x-intercept representing $-1/K_M$.

$$\frac{1}{v} = \frac{K_M}{V_{max}[S]} + \frac{1}{V_{max}}$$

Figure 2.12 shows a Lineweaver–Burk or double-reciprocal plot of kinetic data and the significance of the axis intercepts and slope. The Lineweaver–Burk plot emphasizes the importance of measurements performed at low [S] to obtain good estimations of the V_{max} and K_M values.

Fig. 2.12 Lineweaver–Burk plot

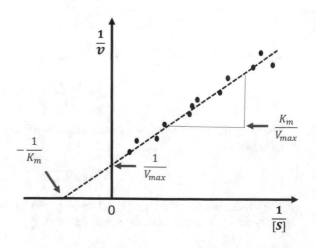

Other linearization forms of the M-M equation have been proposed, such as the Eaddie-Hofstee plot, v versus $v/[S]$, and the Hanes plot, $[S]$ versus $[S]/v$. Alternatively, the kinetic parameters of the M-M equation can be estimated by a nonlinear regression of v versus $[S]$ data.

2.4.2 Multisubstrate Reactions

Multisubstrate reactions may occur via complex rate equations that describe how the substrates bind the enzyme. The study of multisubstrate reactions can be simplified when the concentration of substrate A is kept constant and that of substrate B varies. Under these conditions, the enzyme behaves as a single-substrate enzyme; thus, it is possible to obtain apparent K_M and V_{max} constants as a function of substrate B. Moreover, if a set of experiments is performed with different concentrations of A at a constant B concentration, it is possible to study the reaction mechanism. To obtain a complete description of the kinetics of the reaction A + B $\rightarrow$ P, experiments with different concentrations of B at constant A are also necessary. The experimental design is a matrix of initial rate data at varying concentrations of A and B.

The following are two types of mechanisms by which an enzyme converts two substrate molecules (A and B) into products (P and Q): ternary complex and ping–pong mechanisms.

Ternary complex mechanism

Both substrates bind the enzyme to produce an EAB ternary complex, which is an intermediate of product formation in this type of enzyme-catalysed reaction. As illustrated in Fig. 2.13, the order of binding can be either in a defined sequence (i.e., first A and then B) (ordered mechanism) or random (i.e., either first A and then B or first B and then A) (random mechanism).

Ping-pong mechanism

Enzymes with a ping-pong mechanism can exhibit the following two states: E and a chemically modified form of the enzyme E', namely, a modified enzyme (or intermediate). In this mechanism, substrate A binds and modifies the enzyme (to the E' form), and then, product Y is released. Restoration to the E form occurs by a reaction between E' and substrate B (forming product Z). A schematic of this mechanism is shown in Fig. 2.14. Enzymes with ping–pong mechanisms include certain oxidoreductases (e.g., glucose oxidase), transferases (e.g., acylneuraminate cytidylyltransferase) and serine proteases (e.g., trypsin and chymotrypsin).

For instance, glucose oxidase (GOx) catalyses the oxidation of glucose to hydrogen peroxide and D-glucono-δ-lactone via a ping-pong mechanism. This enzyme can catalyse the oxidation of β-D-glucose into D-glucono-1,5-lactone, which then spontaneously hydrolyses to gluconic acid. To perform the catalytic process, GOx requires a coenzyme, i.e., flavin adenine dinucleotide (FAD^+), which is typically involved in biological oxidation–reduction reactions. In a

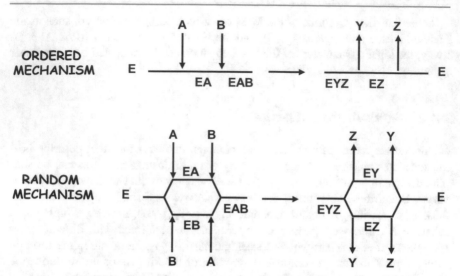

Fig. 2.13 Ordered and random ternary complex mechanisms of an enzyme reaction. The reaction path is shown as a black line, and enzyme intermediates (EA, EB, EAB, EYZ, EY, and EZ) containing substrates (A and B) or products (Y and Z) are indicated below the line. Adapted from Nelson and Cox [2]

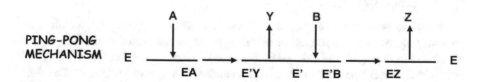

Fig. 2.14 Ping–pong mechanism of an enzyme reaction. Intermediates (enzyme complexes) contain substrates (A and B) or products (P and Q). Adapted from Nelson and Cox [2]

GOx-catalysed redox reaction, FAD^+ acts as the initial electron acceptor, which is reduced to $FADH_2$, while glucose is oxidized to D-glucono-1,5-lactone. Then, $FADH_2$ can be oxidized by molecular oxygen. Finally, in the presence of water, O_2 is reduced to H_2O_2. A schematic of this enzymatic process is shown in Fig. 2.15.

2.4.3 Non-Michaelis–Menten Kinetics

Some enzymes produce a sigmoid (or S-shaped) curve of the reaction rate (v) as a function of [S], which may indicate cooperative binding of the substrate to the active site of the enzyme. Multimeric enzymes with several interacting active sites may exhibit this behaviour [16].

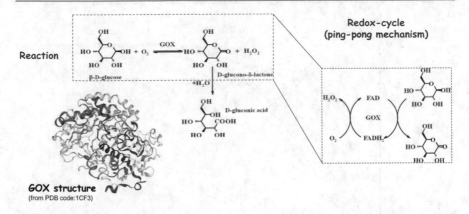

Fig. 2.15 Oxidation of glucose by molecular oxygen into D-glucono-δ-lactone catalysed by glucose oxidase

Overall, positive cooperativity may occur when the binding of the first substrate molecule increases the affinity of the other binding sites for subsequent substrate molecules. In contrast, negative cooperativity can be observed when the binding of the first substrate reduces the affinity between the enzyme and other substrate molecules. For instance, Christian Bohr studied the binding of haemoglobin to oxygen under different conditions [17] and observed a sigmoidal curve when considering haemoglobin saturation with oxygen as a function of the oxygen partial pressure (Fig. 2.16). Therefore, the more oxygen bound to haemoglobin, the easier it is for more oxygen to bind until all binding sites are saturated. Moreover, Bohr observed that increasing the CO_2 pressure shifted this S-shaped curve to the right. In fact, higher CO_2 concentrations render it more difficult for haemoglobin to bind oxygen.

Cooperativity is common in many biochemical reactions and can help regulate the responses of enzymes to variations in [S]. Positive cooperativity renders enzymes much more sensitive to the substrate concentration, and their activities exhibit multiplicative increases over a narrow range of [S]. In contrast, negative cooperativity renders enzymes insensitive to small changes in [S], and the binding of a substrate molecule decreases affinity.

When the plot of fractional occupancy ($\overline{Y}$) is defined as

$$\overline{Y} = \frac{[boundsites]}{[boundsites] + [unboundsites]} = \frac{[boundsites]}{[totalsites]}$$

at equilibrium, a sigmoidal curve is obtained with respect to the substrate concentration. Then, there is positive cooperativity as observed by Bohr with haemoglobin. In contrast, no conclusion can be drawn regarding cooperativity when enzymes do not produce a sigmoidal curve. The concept of cooperative binding ca be applied only to enzymes with more than one binding site. In contrast, if

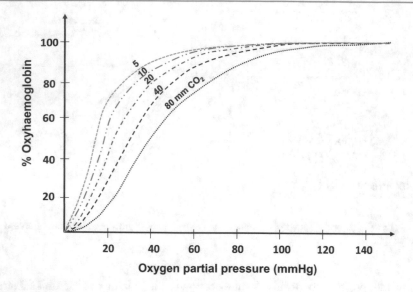

Fig. 2.16 Sigmoidal increase in oxyhaemoglobin production as a function of the partial pressure of oxygen and CO_2. Measurements were performed using freshly obtained dog blood. Adapted from Bohr et al. [17]

n-binding sites exist but substrate binding to any one site does not affect the others, this is noncooperative behaviour.

2.5   Factors Influencing Enzyme Activity

2.5.1   Environmental Conditions: Temperature and pH

Similar to most chemical reactions, the rate of enzyme-catalysed reactions increases as the temperature is raised. In general, the temperature dependence of reaction rate constants can be well described by the Arrhenius equation; even inactivation rates can be described according to Arrhenius-type equations. However, many enzymes are adversely affected by high temperatures. The temperature dependence of enzyme-catalysed reactions exhibits an optimum in terms of activity (usually between 40 and 60 °C) as follows: the thermodynamic increase in the reaction rate to a maximum is followed by a steep decrease due to thermal denaturation of the enzyme. Figure 2.17 shows the temperature dependence of enzyme activity as follows: the reaction rate increases with temperature to a maximum level and then sharply declines with a further increase in temperature [18]. This maximum has a very limited practical meaning since at such temperature, the enzyme is quite unstable; thus, the optimum operation temperature can be substantially lower than this maximum.

Fig. 2.17 Effect of temperature on enzyme-catalysed reactions. Adapted from Aehle et al. [18]

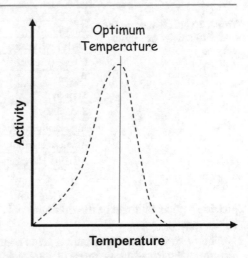

A temperature of 5 °C or below is generally the most suitable condition to preserve enzyme activity during storage. However, some enzymes may lose their activity when frozen. Similarly, enzymes are affected by changes in pH and exhibit an optimum pH for their activity as illustrated in Fig. 2.18. The optimum depends not on only pH but also on the ionic strength and type of buffer.

For most enzymes, the optimum pH is between 5 and 7 [18]. However, the optimum pH can significantly vary across enzymes as shown in Table 2.1.

The extreme pH optima of pepsin and alkaline phosphatase have been observed at pH 1.5 and 10.5, respectively.

Notably, in contrast to the effect of temperature, in the case of pH, there is no compromise between activity and stability. In most cases, the pH at which activity is maximized is within the region of higher stability. This finding is important because the pH-activity profile (which is quite straightforward to determine) can

Fig. 2.18 Effect of pH on the activity of pepsin, saccharase and trypsin. Adapted from Aehle et al. [18]

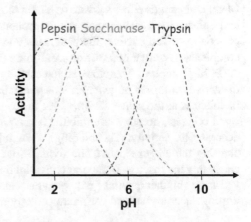

Table 2.1 pH ranges for the optimum activity of various enzyme

Enzyme	Optimum pH
Lipase (pancreas)	8.0
Lipase (stomach)	4.0–5.0
Pepsin	1.5
Trypsin	7.8–8.7
Saccharase	4.5
Amylase (pancreas)	6.7–7.0
Amylase (malt)	4.6–5.2
Alkaline phosphatase	10.5

provide a good estimate of the optimum pH at which the enzymatic reaction should be conducted.

The presence of charged amino acids contributes to the net surface charge of an enzyme, which can be positive or negative depending on the pH of the medium. The pH at which the protein has no net charge is defined as the isoelectric point (pI) and provides information regarding the binding between enzymes and ionic supports in immobilized enzymes. At extreme pH values, the cationic or anionic characteristic of an enzyme is favoured, thus promoting interaction with the solid support. However, extreme pH conditions can inactivate the enzyme. Thus, as a general rule, enzymes should be 0.5–1 pH units above or below their pI to bind a positively or negatively charged matrix, respectively.

The effect of pH on the kinetic parameters (K_M and V_{max}) can be estimated from the Michaelis-Davidsohn hypothesis [19]. In fact, explicit functions in terms of the hydrogen ion concentration can be obtained from the assumptions underlying this hypothesis and determined using experimental data of reaction rates at different pH values.

2.5.2   Activators and Inhibitors

Many compounds either activate or inhibit enzyme activity. In addition to coenzymes and substrates, enzymes may require nonprotein or, in some cases, protein compounds to be fully active. Enzyme activation by inorganic ions can be achieved by complexation with a cofactor or cosubstrate (e.g., Fe binding bound to flavin or the ATP–Mg complex). Therefore, an ion can be a part of an enzyme that either stabilizes the active conformation (e.g., Zn ions in alkaline phosphatase) or directly participates in catalysis at the active site (e.g., Zn or Co ions in carboxypeptidases) [18].

In contrast, an enzyme inhibitor is a compound that may bind an enzyme and decrease its activity. The binding of an inhibitor can prevent a substrate from entering the enzyme active site (competitive inhibition), hinder the enzyme from catalysing the reaction (noncompetitive inhibition), or both (mixed-type inhibition).

Many biotherapeutics are enzyme inhibitors; thus, their discovery and improvement are critical for pharmacology research.

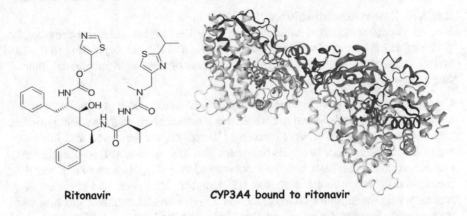

Ritonavir   CYP3A4 bound to ritonavir

Fig. 2.19 Structure of human CYP3A4 bound to ritonavir from PDB code 5VC0 [20]

For example, ritonavir was originally developed as a human immunodeficiency virus (HIV) protease inhibitor. The molecular structure of this complex enzyme inhibitor is shown in Fig. 2.19. Although this compound is currently rarely used for its antiviral activity, it remains widely used as a booster of protease inhibitors. In fact, ritonavir inhibits the enzyme cytochrome P450-3A4 (CYP3A4), which is an important enzyme mainly found in the liver and the intestine. CYP3A4 oxidizes small foreign organic molecules, such as toxins or drugs, allowing them to be removed from the body.

However, the mechanism by which ritonavir inhibits CYP3A4 remains unclear. Recent data suggest that multiple types of inhibition occur, including mechanism-based inactivation by metabolic-intermediate complex formation, competitive inhibition, irreversible type II coordination to haem iron, and haem destruction. Moreover, it has been demonstrated that the inhibition of CYP3A4 by ritonavir occurs by CYP3A4-mediated activation and the subsequent formation of a covalent bond with apoprotein [21].

From a biocatalysis perspective, enzyme inhibition is quite important since many enzymes are inhibited by the products of the reaction and even the substrate at high concentrations.

Enzyme inhibition can be either reversible or irreversible. Depending on the type of inhibitory effect, the following mechanisms of enzyme inhibition may be distinguished:

(i) Reversible inhibitors bind noncovalently, and the inhibitors may bind the enzyme, the enzyme–substrate complex, or both.

(ii) Irreversible inhibitors typically change the enzyme chemically (e.g., by forming covalent bonds). These inhibitors modify the amino acid residues needed for enzyme activity.

2.5.2.1 Reversible Inhibition

Since the inhibitor can bind either the enzyme or the enzyme–substrate complex, it is distinguished by two dissociation constants, K_I and K_I', corresponding to the EI and ESI complexes, respectively. The mechanisms of reversible enzyme inhibition can be classified as follows:

Competitive inhibition: This form of inhibition results from an inhibitor with affinity for an enzyme's active site where the substrate binds such that the inhibitor and substrate cannot bind simultaneously. Therefore, the substrate and inhibitor compete with each other to access the active site. This particular type of inhibition can be overcome at high substrate concentrations. V_{max} is constant since the inhibitor does not interact with the ES complex. However, the apparent K_M increases with the inhibitor concentration as a higher concentration of the substrate is required to prevent the nonproductive binding of the inhibitor to the active site. Competitive inhibitors are often structurally similar to the substrate. As shown in Fig. 2.20, competitive inhibitors bind E but not ES.

Noncompetitive inhibition: The binding of an inhibitor to an enzyme reduces its activity but the binding of the substrate to the active site is not affected. In fact, a non-competitive inhibitor binds the enzyme away from the active site, changing the conformation of the enzyme such that its active site no longer operates (total non-competitive) or functions at a lower rate (partial non-competitive). These inhibitors bind allosteric sites and hamper enzyme activity by modifying the structure of the active site (disrupting the normal arrangement of the H-bonds and weak hydrophobic interactions keeping the enzyme molecule together in its 3-D shape). The degree of inhibition depends on the inhibitor concentration. Affinities for E and ES ($K_I = K_I'$) in non-competitive inhibitors are identical; thus, binding the inhibitor to its site does not affect the binding of the substrate to the active site and

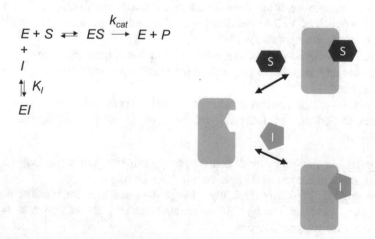

$$E + S \rightleftharpoons ES \xrightarrow{k_{cat}} E + P$$

Fig. 2.20 Competitive inhibition. Adapted from Nelson and Cox [2]

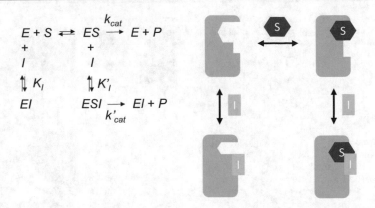

$$E + S \rightleftharpoons ES \xrightarrow{k_{cat}} E + P$$

Fig. 2.21 Noncompetitive inhibition. Adapted from Nelson and Cox [2]

vice versa. V_{max} decreases because it is impossible or difficult for the reaction to proceed, but K_M stays the same since the substrate can still properly bind. The presence of the inhibitor modifies the enzyme structure and either limits the interaction of the enzyme and the substrate or hampers the conversion of the bound substrate into the final product (Fig. 2.21). Partial noncompetitive inhibition may occur when the ESI complex yields a product but at a lower rate ($k'_{cat} < k_{cat}$) than the ES complex.

Uncompetitive inhibition: The inhibitor only binds the ES complex. This type of inhibition decreases both V_{max} (as a consequence of removing the activated complex) and K_M. The decrease in K_M is a result of Le Chatelier's principle (better binding efficiency) and the effective elimination of the ES complex. Actually, the effects of an uncompetitive inhibitor on V_{max} and K_M are of the same magnitude. Figure 2.22 shows a scheme of this inhibition mechanism.

Mixed inhibition: Similar to non-competitive inhibition, the inhibitor can bind the enzyme at the same time as the substrate. Though mixed-type inhibitors can bind the active site, this type of inhibition is generally a consequence of an allosteric effect in which the inhibitor binds a site other than the active site of the enzyme. An inhibitor binding this allosteric site modifies the structural conformation of the enzyme, such that there is a reduction in the affinity of the substrate for the active site. A schematic is shown in Fig. 2.23.

Mixed-type inhibitors bind the enzyme as well as the enzyme–substrate complex. However, their affinities for these two types of the enzyme differ ($K_I \neq K_I'$). In fact, non-competitive inhibition is a particular case of mixed inhibition (namely, when $K_I = K_I'$).

The ESI complex can yield products but at a lower rate ($k'_{cat} < k_{cat}$) than the ES complex.

It is possible to measure the enzyme-inhibitor constant K_I by several techniques, including isothermal titration calorimetry, where the inhibitor is titrated into an

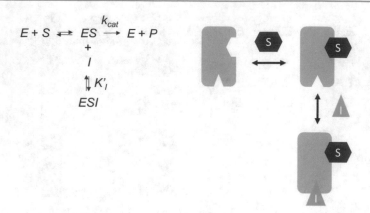

Fig. 2.22 Uncompetitive inhibition. Adapted from Nelson and Cox [2]

$$E + S \;\rightleftharpoons\; ES \;\xrightarrow{k_{cat}}\; E + P$$
$$+$$
$$I$$
$$\updownarrow K'_I$$
$$ESI$$

$$
\begin{array}{ccc}
E + S & \rightleftharpoons & ES \xrightarrow{k_{cat}} E + P \\
+ & & + \\
I & & I \\
\updownarrow K_I & & \updownarrow K'_I \\
EI + S & \rightleftharpoons & ESI \xrightarrow[k'_{cat}]{} EI + P
\end{array}
$$

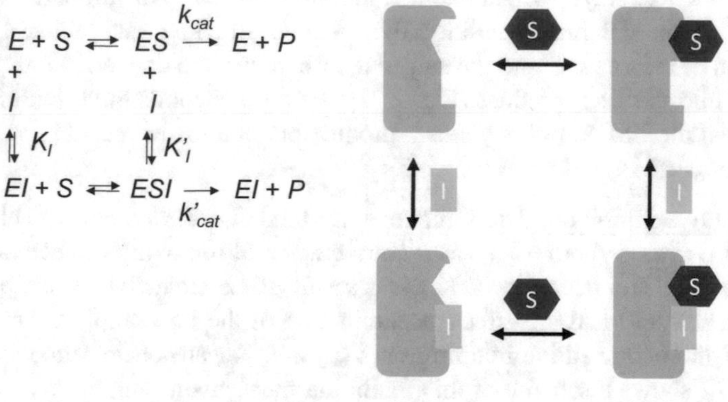

Fig. 2.23 Mixed inhibition. Adapted from Nelson and Cox [2]

enzyme solution and the heat that is released or absorbed is measured. However, it is difficult to directly measure the K_I', as the enzyme–substrate complex is short-lived and undergoes a chemical reaction to form the product. K_I' can be estimated indirectly though, by determining the enzyme activity at various substrate and inhibitor concentrations and by fitting the data to the following modified Michaelis–Menten equation:

$$v = \frac{V_{\max}[S]}{\alpha K_M + \alpha'[S]} = \frac{(1/\alpha')V_{\max}[S]}{(\alpha/\alpha')K_M + [S]}$$

where the modifying factors α and α' are as follows:

$$\alpha = 1 + \frac{[I]}{K_I}$$

$$\alpha' = 1 + \frac{[I]}{K_I'}$$

In the presence of reversible inhibitor I, the effective K_M and V_{max} of an enzyme become $(\alpha/\alpha')K_M$ and $(1/\alpha')V_{max}$, respectively.

The effects of different forms of reversible enzyme inhibitors on enzymatic activity may be analysed using linearized forms of the Michaelis–Menten equation, such as Lineweaver–Burk plots. As shown in Fig. 2.24, the competitive inhibition lines intersect on the y-axis, demonstrating that such inhibitors do not affect V_{max}. Likewise, the non-competitive inhibition lines intersect on the x-axis, illustrating that K_M is not affected by these inhibitors. However, the Lineweaver–Burk plot of uncompetitive inhibition yields parallel rather than intersecting lines, showing that the effects on V_{max} and K_M are of the same magnitude. These effects occur because there is a reduction in the concentration of the ES complex, which decreases the maximum enzyme activity (V_{max}) because it takes longer for molecules to leave the active site. Moreover, there is a reduction in K_M since the inhibitor bound to the ES complex reduces the concentration of the complex in solution. Finally, mixed-type inhibitors both interfere with substrate binding (increase K_M) and obstruct catalytic activity in the ES complex (decrease V_{max}). This finding is revealed in a Lineweaver–Burk plot by the intersection of lines at a point away from both the y- and x-axes located in quadrant II. Notably, in mixed inhibition, some cross-situations may occur, namely, a decrease in both V_{max} and K_M or an increase in both V_{max} and K_M, even though these mechanisms are less frequent.

Substrate Inhibition: High substrate (or coenzyme) concentrations can bring about a decrease in the catalytic activity of an enzyme. Such an effect normally occurs at high substrate concentrations and is a result of a substrate binding a form of an enzyme to which the product of the substrate usually combines. Inhibition at high substrate concentrations, which is not infrequent, is usually described by an uncompetitive inhibition mechanism in which a second nonproductive binding site for the substrate exists when the enzyme is already saturated with the substrate at the active site. In this case, the resulting rate equation is non-Michaelian, and a parabolic rather than hyperbolic curve is obtained for v versus [S]. In this case, linearization in the Lineweaver–Burk plot is possible only at very low or very high [S], which makes determining the kinetic parameters quite prone to error. In this case, it is advisable to determine the kinetic parameters by a nonlinear regression of experimental data to the rate equation. Substrate inhibition is easily visualized in a v versus [S] plot, where a maximum v is obtained at a finite [S]. The optimum [S] (the value that maximizes v) corresponds to the root square of the product of the M-M constant and the substrate inhibition constant.

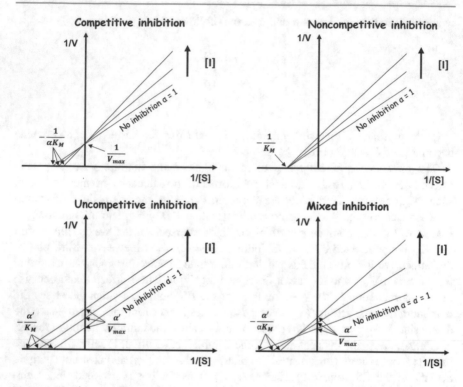

Fig. 2.24 Effects of different types of reversible enzyme inhibitors on reaction rates. Adapted from Nelson and Cox [2]

Feedback end-product inhibition: In several multienzyme systems, the reaction sequence end product may inhibit an enzyme at or close to the beginning of the process. In other words, the rate of the whole reaction sequence can be determined by the steady-state concentration of the end product. This type of inhibition is an important mechanism of metabolic control.

2.5.2.2 Irreversible Inhibition

An irreversible inhibitor usually forms a stable complex with an enzyme through covalent bonding with an amino acid residue at the enzyme active site. Figure 2.25 shows an interaction between diisopropylfluorophosphate (DFP) and a serine protease leading to irreversible inhibition. Irreversible inhibition differs from irreversible enzyme inactivation since irreversible inhibitors are usually specific to one class of enzymes and do not inactivate other proteins. These inhibitors do not destroy the protein structure and function by specifically altering the active site. In contrast, extreme conditions, such as extreme pH, high temperatures or enzyme poisons, usually cause nonspecific denaturation of the protein structure. Similarly, certain chemicals may destroy the protein structure, thereby causing complete enzyme inactivation.

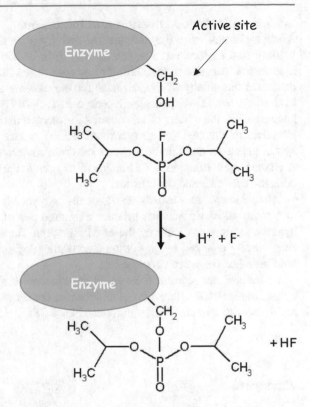

Fig. 2.25 Interaction between diisopropylfluorophosphate (DFP) and a serine protease. Adapted from Nelson and Cox [2]

As shown in Fig. 2.26, irreversible inhibitors form a reversible complex with the enzyme, namely, EI or ESI, which, in turn, reacts to form the covalently modified "dead-end complex" (EI*). The reaction rate constant at which EI* is formed is designated K_{inact}. The formation of the EI complex may compete with the formation of ES; thus, the binding of irreversible inhibitors can be prevented by favourable competition at high substrate concentrations.

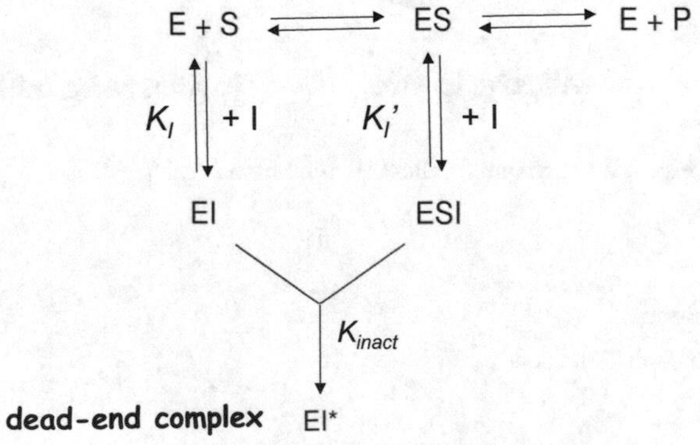

Fig. 2.26 Kinetic scheme of irreversible inhibition

2.5.2.3 Allosteric Activation and Inhibition

Allostery occurs when the structure and activity of an enzyme are modified by the binding of a molecule at a region distinct from its active site. Negative allosteric modulation (or allosteric inhibition) occurs when the binding of one ligand decreases the affinity of the substrate for the enzyme active sites (Fig. 2.27). For example, when 2,3-bisphosphoglyceric acid (2,3-BPG) binds an allosteric site of haemoglobin, the affinity of all subunits for oxygen decreases. In contrast, positive allosteric modulation (or allosteric activation) occurs when the binding of one ligand enhances the affinity between substrate molecules and other binding sites. A characteristic example is the binding of oxygen to haemoglobin, where oxygen is both the substrate and the effector.

Allosteric enzymes usually do not exhibit typical Michaelis–Menten kinetics. In fact, many allosteric enzymes produce a sigmoid plot of v versus [S] rather than the hyperbolic plots predicted by the M-M equation. The shape of the curve is characteristically changed by an allosteric activator (positive cooperation) or an allosteric inhibitor (negative cooperation).

In contrast, the sigmoidal curve (in the absence of allosteric activation or inhibition) implies that within a range of substrate concentrations, enzymatic activity is highly sensitive to small [S] variations (Fig. 2.28).

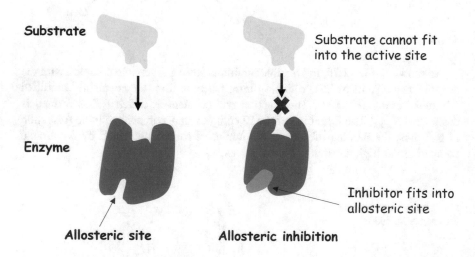

Fig. 2.27 Scheme of allosteric inhibition

Fig. 2.28 Effects of allosteric activation and inhibition on the reaction rate

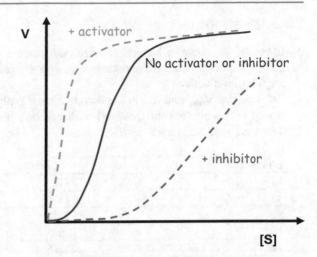

2.6 Summary

- Enzymes increase reaction rates by 10^8 to 10^{20} times and are extremely active due to their substrate complementarity, which must be considered in both electrostatic and geometric terms.
- Enzymes convert substrates into products via multistep reaction mechanisms (enzymatic cycles). Similar to other catalysts, enzymes do not modify the equilibrium position between substrates and products but merely increase the rate at which this equilibrium is approached. However, enzyme-catalysed reactions typically exhibit saturation kinetics (e.g., Michaelis–Menten kinetics).
- Although the mechanism of an enzyme reaction can be quite complex, there is typically one rate-determining step, which allows the reaction to be modelled as a single catalytic step.
- Multimeric enzymes may exhibit cooperative behaviours.
- Enzyme inhibition can be either reversible or irreversible as follows: reversible inhibitors bind noncovalently, and different types of inhibition are possible (e.g., competitive, uncompetitive, noncompetitive, and mixed-type), while irreversible inhibitors typically interact with and change the enzyme chemically (e.g., via covalent bonds).

2.7 Questions

1. Enzymes are effective biocatalysts that can increase reaction rates by 10^8 and 10^{20} times and at mild temperatures. Can you explain the fundamental reasons for their high activity?

2. Estimate the V_{max} and K_M in the conversion of p-nitrophenyl phosphate (substrate) into p-nitrophenol (product) with alkaline phosphatase. The following data were obtained experimentally:

[S] (μM)	V (μM/s)
16	310
40	1325
80	1718
160	2683
320	4216
640	6000
1280	6167

3. In which type of inhibition do both V_{max} and K_M decrease? Why?

4. What would happen to haemoglobin in the presence of carbon monoxide? How does carbon dioxide affect haemoglobin?

5. Some enzymes exhibit a sigmoid curve of the reaction rate as a function of the substrate concentration rather than the hyperbolic plots predicted by the Michaelis–Menten equation. Give some examples and explain why these behaviours occur.

References

1. N.-C. Ha, S.-T. Oh, J.Y. Sung, K.-A. Cha, M. Hyung Lee, B.-H. Oh, Nat. Struct. Mol. Biol. **8**, 505–509 (2001)
2. D.L. Nelson, M.M. Cox, *Lehninger Principles of Biochemistry*, 7th edn. (MacMillan, 2017), pp. 187–235
3. D.C. Phillips, PNAS **57**(3), 483–495 (1967)
4. E. Fischer, Ber. Dtsch. Chem. Ges. **23**, 2611 (1890)
5. E. Fischer, Ber. Dtsch. Chem. Ges. **27**, 2985 (1894)
6. D.E. Koshland, Angew Chem. Int. Ed. Engl. **33**, 2375–2378 (1994)
7. A. Fersht, *Enzyme Structure and Mechanism* (W.H. Freeman, San Francisco, 1985), pp. 50–52
8. A. Warshel, P.K. Sharma, M. Kato, Y. Xiang, H. Liu, M.H. Olsson, Chem. Rev. **106**(8), 3210–3235 (2006)
9. S.J. Benkovis, S. Hammes-Schiffer, Science **301**(5637),1196–202 (2003)
10. J. Villa, M. Strajbl, T.M. Glennon, Y.Y. Sham, Z.T. Chu, A. Warshel A, PNAS **97**(22), 11899–904 (2000)
11. L. Polgár, Cell. Mol. Life Sci. **62**, 2161–2172 (2005)
12. E.Z. Eisenmesser, D.A. Bosco, M. Akke, D. Kern, Science **295**(5559), 1520–1523 (2002)
13. J.B. Murray, C.M. Dunham, W.G. Scott. J. Mol. Biol. **315**(2), 121–30 (2002)

14. L. Michaelis, M.L. Menten, Biochem Z. **49**, 333–369 (1913)
15. G.E. Briggs, J.B.S. Haldane, Biochem J. **19**, 338–339 (1925)
16. J. Ricard, A. Cornish-Bowden, Eur. J. Biochem. **166**(2), 255–272 (1987)
17. C. Bohr, K. Hasselbalch, A. Krogh, Skandinavisches Archiv Für Physiologie **16**(2), 402–412 (1904)
18. W. Aehle, N.R. Perham, G. Michal, A.J. Caddow, B. Concoby, Enzymes, in *Ullmann's Encyclopedia of Industrial Chemistry* (Wiley-VCH Verlag GmbH & Co. KGaA, Weinheim, 2012), pp. 573–611
19. A. Illanes, *Enzyme Biocatalysis* (Springer, Netherlands, 2008)
20. I.F. Sevrioukova, Biochemistry **56**, 3058–3067 (2017)
21. B.M. Rock, S.M. Hengel, D.A. Rock, L.C. Wienkers, K.L. Kunze, Mol Pharmacol. **86**(6), 665–674 (2014)

Introduction to Molecular Catalysis

3

Marco Piumetti and Andrés Illanes

3.1 Science of Catalysis [1–3]

Empirical studies investigating the natural world have been described since classical antiquity, and alchemy was practised by several civilizations for many centuries. For approximately 10,000 years, humanity believed that chemical changes occurred for mysterious reasons, sometimes produced by acts of gods. It was only at the beginning of the nineteenth century that the chemist Jöns J. Berzelius coined the terms "catalysis" (from "loosening down" in Greek) and "catalytic force".

In a review in 1835, Berzelius considered several reactions that occurred in the presence of a specific substance that remained unaffected; he wrote the following: "In order to avail myself of a derivation well known in chemistry, I will call both the catalytic force of matter and the decomposition by this matter, catalysis, just as we understand by the word analysis the separation of the constituents". Berzelius endowed catalysts with a mysterious property called catalytic force, and this aura of mystery was described from a scientific perspective only in relatively recent times. In 1895, Wilhelm Ostwald defined catalysis as the acceleration of chemical reactions by the presence of foreign substances that are not consumed. Similar to Liebig before him, Ostwald described the catalytic process with the following analogy: "a catalyst acts like oil on a machine, or a whip on a tired horse". The current definition of a catalyst, which was originally proposed by Ostwald, states that a catalyst is a substance that changes the rate but not the thermodynamics of a chemical reaction. According to IUPAC (1997), a catalyst is a substance that increases the reaction rate without modifying the overall standard Gibbs energy change in the reaction. Therefore, a catalyst can be described as a substance that increases the rate of attainment of chemical equilibrium without undergoing a chemical change and without altering that equilibrium. A catalytic reaction is a

Electronic Supplementary Material The online version of this chapter (https://doi.org/10.1007/978-3-030-88500-7_3) contains supplementary material, which is available to authorized users.

cyclic event in which a catalyst participates and is recovered in its "original form" at the end of the cycle.

3.1.1 Homogeneous, Heterogeneous and Enzymatic Catalysis

Overall, it is possible to distinguish among several catalytic domains as follows: homogeneous (or molecular) catalysis occurs when the catalyst and reactants are in the same medium (either liquid or gaseous); heterogeneous catalysis, including photocatalysis and electrocatalysis, occurs when the catalyst and reactants are in different phases; and enzymatic catalysis (or biocatalysis) may exhibit both homogeneous and heterogeneous characteristics.

The solubility criterion, namely, the presence of a catalyst and reactants in one or more phases, is conventionally used to categorize "homogeneous" and "heterogeneous" systems. Homogeneous catalysis involves processes that occur entirely in a common phase (i.e., all molecules are in a liquid or a gas phase). Homogeneous catalysts generally comprise individual molecules exhibiting single-site and molecular characteristics. Metal salts of organic acids, organometallic complexes, and carbonyls of Co, Fe, and Rh are typical homogeneous catalysts. An example is the catalytic carbonylation of methanol to acetic acid as follows:

$$CH_3OH + CO \rightarrow CH_3COOH$$

This reaction is performed using $[Rh(CO)_2I_2]^-$ complexes dissolved in solution. Another characteristic example is the Wacker process, which involves the conversion of ethene into acetaldehyde by oxygen in water in the presence of Pd^{2+} and Cu^{2+} (redox-type mechanism) as follows:

$$C_2H_4 + {}^1\!/_2O_2 \rightarrow CH_3CHO$$
$$C_2H_4 + PdCl_2 + H_2O \rightarrow CH_3CHO + Pd^0 + 2HCl$$
$$2CuCl_2 + Pd^0 \rightarrow 2CuCl + PdCl_2$$
$$2CuCl + 2HCl + {}^1\!/_2O_2 \rightarrow 2CuCl_2 + H_2O$$
$$-\,-$$
$$2C_2H_4 + O_2 \rightarrow 2CH_3CHO$$

The advantages of homogeneous catalysis include the following [4]:

- High selectivity and reproducibility
- Mild reaction conditions
- Easy to study the reaction mechanisms
- Great efficiency because all atoms participate in catalysis.

However, homogeneous catalysts have the following practical disadvantages:

- High cost
- Poor robustness
- Short lifespan.

In addition to these disadvantages, homogeneous catalysis often occurs in environmentally aggressive solvents.

Table 3.1 summarizes the main advantages and disadvantages of homogeneous catalysis.

In contrast, in heterogeneous catalysis, the catalyst and reactants are in separate physical phases. Thus, the reaction occurs at the interface between two phases. Solids commonly catalyse reactions of molecules in the gas phase (solid–gas systems) or liquid phase (solid–liquid systems). The typical heterogeneous catalysts include inorganic solids, such as metals, oxides, sulfides and metal salts, although they may also include organic materials, such as organic hydroperoxides and ion exchangers. Solid catalysts frequently involve reactions that occur over a broad temperature range (up to 1000 °C) and have to be much more thermally stable than the catalysts in homogeneous systems, which typically occur in the 50–200 °C temperature range and even at lower temperatures in the case of enzymes. Regarding separation processes, solid catalysts offer a major advantage; in fact, heterogeneous catalysts may be easily separated from the reaction mixture to be recycled and reused until the catalytic activity is reduced to the point of replacement. Moreover, these catalysts can be easily adapted for continuous processes. Thus, many large-scale processes, such as the conversion of chemicals, fuels and pollutants, occur via heterogeneous catalysis. Heterogeneous catalytic reactions occur at solid surfaces and can be accelerated by increasing the surface area. Many elements and compounds, including metals, metal oxides and metal sulfides, are active as catalysts for a wide number of reactions. However, few solid catalysts used in industry have simple compositions (e.g., metals or metal oxides). In fact, industrial catalysts are usually complex multicomponent solids that often require an

Table 3.1 Main advantages and disadvantages of homogeneous catalysis [1, 3]

Advantages	Disadvantages
– High selectivity and reproducibility – Efficient utilization of active sites	– Difficult catalyst separation and recycling (high costs for recycling and regeneration)
– Well-defined and accessible active sites that can be characterized in situ with high certainty	– Solvent (or liquid medium) is required: product recovery can be difficult and expensive – Usually, environmentally aggressive solvents are used
– The reaction pathway can be understood	– Sometimes they exhibit high corrosion
– Systems are active at a relatively low temperature and pressure	– Sensitive systems

activation procedure, such as some time on stream (TOS). Thus, catalyst preparation may involve different catalyst states that are a function of the oxidation degree, water content, and bulk structure and have different kinetic properties (activity and selectivity).

However, the main disadvantages of heterogeneous catalysts are the following:

- The active sites are normally unknown.
- The reproducible synthesis of empirically derived catalyst formulations can be difficult.
- Molecular modelling for rational design is often a difficult task.

Table 3.2 shows the comparative advantages and disadvantages of heterogeneous and homogeneous catalysts; three main properties are considered, namely, catalyst recovery, selectivity and thermal stability.

In recent years, the boundaries between homogeneous and heterogeneous catalysis have become less defined, and there has been increased convergence between these two fields. In particular, interest in catalysis using metal nanoparticles (NPs) has markedly increased. This field, which is also known as "semi-heterogeneous" catalysis, is at the frontier between homogeneous and heterogeneous catalysis. Enzymes have features of both homogenous and heterogeneous catalysts. Similar to heterogeneous catalysts, enzymes have an active surface on which a reactant is immobilized temporarily while waiting for the subsequent steps. Similar to homogenous catalysts, the amino acid groups of enzymes actively interact with substrates in multistep sequences involving several intermediates. Enzymatic catalysis is molecular, single-site and not based on physical properties unique to bulk aggregations of atoms as observed in heterogeneous catalysis [4]. Thus, enzymes can be either homogeneous or heterogeneous catalysts depending on whether the enzyme is dissolved in the reaction medium or immobilized in a solid structure.

3.1.2   Catalytic Activity, Selectivity and Yield

Catalytic activity is defined in terms of the initial reaction rates, preferably normalized to the surface area of the active phase (areal rates), which can be measured

Table 3.2 Comparison of some properties of homogeneous and heterogeneous catalysts [1, 3]

Property	Homogeneous catalysts	Heterogeneous catalysts
Catalyst recovery	Difficult separation and recycling (costly procedure)	Easy and inexpensive
Selectivity	Excellent/good	Good/low
Thermal stability	Low	High

by chemisorption. Alternatively, the specific reaction rate, i.e., the reaction rate normalized by the catalyst weight, can be used. The best possible measure of catalytic activity is the turnover rate or turnover frequency (TOF; unit s^{-1}). The TOF is defined as the number of times (n) that the overall catalytic reaction occurs per catalytic site per unit time under specific reaction conditions (e.g., temperature, pressure, concentration, and reactant ratio).

Therefore, the TOF can be written as follows:

$$TOF[s^{-1}] = \frac{number\ of\ molecules\ of\ a\ given\ product}{number\ of\ active\ sites \times time}$$

For most relevant industrial applications, the TOF is in the range from 10^{-2} to $10^2\ s^{-1}$, whereas for enzymes, the TOF ranges between 10^3 and $10^7\ s^{-1}$. Carbonic anhydrase is among the most well-known active catalysts; one molecule of carbonic anhydrase converts 600,000 CO_2 molecules per second (in our muscles) into aqueous carbonic acid (in our bloodstream) [4].

Since the number of active sites for enzymatic and homogeneous processes is usually known, the TOF can be accurately measured. However, in heterogeneous catalysis, it is often difficult to evaluate the number and size of active sites. In such situations, the number of active sites can be replaced by the unit total area of the exposed catalyst. Therefore, it is assumed that the number of sites is proportional to the specific surface area. Thus, the TOF expressed per unit total area must be referred to as the "areal rate of reaction". Similarly, the TOF can be expressed per unit mass or per unit volume of catalyst. However, there are cases (e.g., alloys or multicomponent solids) in which sets of atoms can be distinguished, which allows us to determine the number of active sites in a well-defined structure. For example, in butane oxidation to maleic anhydride over $(VO)_2P_2O_7$-based catalysts, it has been proposed that the active sites consist of four $(VO)_2P_2O_7$ dimer sites separated from one another by excess surface P entities [5].

Additionally, conversion alone is insufficient as a measure of catalytic activity. In fact, the percentage conversion (C, unit %) of inlet feed (e.g., reactant A) can be calculated as follows:

$$C[\%] = \frac{moles_{A,converted}}{moles_{A,fresh}} \times 100$$

However, conversion depends on the surface area of the catalyst and the TOS. The most important property of a catalyst is probably its selectivity (S), which represents its ability to ensure the conversion of reactants along a specific pathway. Selectivity reflects the relative rates of two or more competing chemical reactions. Thus, selectivity is defined as the ratio of the rate of formation of the desired product to the rate of consumption of the starting material. For example, when one reactant (A) is transformed into several products in parallel (B and C) at rates r_1 and r_2, respectively (Scheme 3.1), the selectivity for B can be calculated as follows:

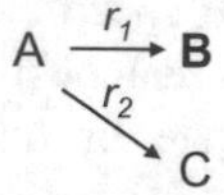

$$A \xrightarrow{r_1} B$$
$$\searrow_{r_2}$$
$$C$$

Scheme 3.1 A reactant (A) transforms into products via parallel reactions (A → B and A → C) [3]

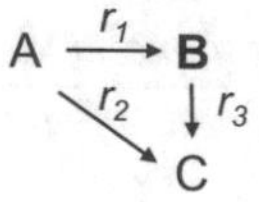

$$A \xrightarrow{r_1} B$$
$$\searrow_{r_2} \quad \downarrow r_3$$
$$C$$

Scheme 3.2 A reactant (A) transforms into products via parallel (A → B and A → C) and consecutive reactions (B → C) [3]

$$S = \frac{r_1}{r_1 + r_2}$$

However, if product B is transformed into C at rate r_3 (Scheme 3.2), selectivity is defined as follows:

$$S = \frac{r_1 - r_3}{r_1 + r_2}$$

Selectivity is determined by the temperature, pressure and molar fractions of the reactants. To compare selectivities, the conversion values must be considered (e.g., these values must be kept constant). Selectivity depends on the conversion of the reactants, which is variable over time. For example, in consecutive reactions (Scheme 3.3), the selectivity for B is initially 100% and then gradually decreases with increasing conversion.

Another measure of selectivity is yield. The relationship among yield (Y), conversion (C) and selectivity (S) is as follows:

$$Y = S \times C$$

The concept of yield enables a comparison of industrial catalysts since both the conversion and selectivity are considered. Yield is typically reported as a percentage and reflects the amount of final product as a percentage of the theoretical yield of the final product (according to the reaction stoichiometry).

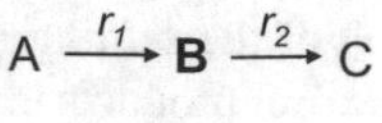

$$A \xrightarrow{r_1} B \xrightarrow{r_2} C$$

Scheme 3.3 Reactant (A) transforms into products B and C via consecutive reactions [3]

3.2 Kinetics of Catalytic Reactions [1, 6–8]

In heterogeneous catalysis, a reaction occurs via a sequence of elementary reaction steps, including adsorption, surface diffusion, chemical transformation of the adsorbed species and desorption. Moreover, the active sites on a solid catalyst must be regenerated after each reaction cycle (self-repair). These elementary reaction steps and surface intermediates constitute the catalytic cycle (Fig. 3.1), and their rates provide useful guidelines for the development of activity and selectivity. The catalytic cycle provides the best description of a catalysed reaction at the molecular scale.

The time scales of the adsorption, desorption, diffusion and reaction steps are often 10^{-3} s or less, whereas the movement of surface atoms can be on the order of 10^{-9} s. The vibrational frequencies of adsorbed species occur at time scales on the order of 10^{-12} s. Therefore, catalysts operate through sophisticated phenomena occurring over different time and spatial scales. Similar to heterogeneous catalysis, in both homogeneous and enzymatic catalysis, several events occur in the catalytic cycle. The reaction occurs via intermediate species, and the catalyst is almost completely renewed after each cycle via self-repair phenomena.

At the macroscopic scale, the rate of a catalytic reaction can be modelled by fitting empirical equations to experimental data; hence, it is possible to determine the kinetic parameters. This approach is useful for reactor and process design in chemical engineering. Additionally, a catalytic reaction can be described using a microkinetics approach by modelling the macroscopic kinetics by correlating

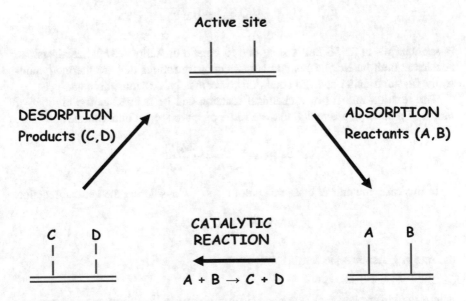

Fig. 3.1 Catalytic cycle of the catalytic reaction A + B → C + D. Adapted from Piumetti and Russo [3]

atomic processes and macroscopic parameters (e.g., partial pressures and temperature) within the framework of a continuum model. The kinetic model is based on an accurate description of the catalytic cycle (in terms of information regarding both the elementary steps and the active sites).

3.2.1 Reaction Rates

Thermodynamics enables the prediction of whether a chemical reaction is spontaneous and can be used to evaluate the energy release/uptake during the reaction. The equilibrium position reached in a catalysed reaction is the same as that obtained without a catalyst. The equilibrium constant (K) depends on the Gibbs free energy change of the reaction, which, in turn, is determined by the changes in enthalpy and entropy as follows:

$$\Delta G = \Delta H - T\Delta S$$
$$K = \exp(-\Delta G/RT)$$

Thus, the spontaneity of a reaction can be predicted using a parameter that considers both ΔH and ΔS as follows: when ΔG is negative, the reaction is favoured and releases energy. The Gibbs free energy is also the chemical potential that is minimized when a system reaches equilibrium at constant pressure and temperature. However, some reactions that are thermodynamically favourable ($\Delta G < 0$) occur very slowly. For example, the reaction

$$H_{2(g)} + \tfrac{1}{2}O_{2(g)} \rightarrow H_2O_{(l)}$$

is spontaneous at 25 °C but is slow due to kinetic limitations. Thus, catalysts can accelerate both heterogeneous and homogeneous reactions that are thermodynamically favoured under specific conditions; specifically, ΔG must be negative.

The reaction rate (v) of a chemical reaction can be defined as the number of moles of product(s)/reactant(s) that are formed/converted per unit volume and time.

$$A \rightarrow Bv = -\frac{\Delta[A]}{\Delta t} = \frac{\Delta[B]}{\Delta t}$$

In this case, the unit of measure is mol l^{-1} s^{-1}. Considering the general reaction

$$aA + bB \rightarrow cC + dD,$$

the rate law assumes the form.
$rate = k[A]^x[B]^y,$

where the kinetic constant (k) depends on the temperature, reactants and catalyst surface (when a heterogeneous catalyst is involved), and x and y are exponents that must be determined experimentally. The exponents x and y specify the relationships between the concentrations of reactants A and B and the reaction rate and do not necessarily correspond to the stoichiometric coefficients a and b. When added, $x + y$ provides the overall reaction order defined as the sum of the powers to which all reactant concentrations appearing in the rate law are raised. The reaction order is always defined in terms of the reactant (not product) concentrations.

For example, for the general reaction $aA + bB \rightarrow cC + dD$, $x = 1$ and $y = 3$. Thus, the rate law is as follows:

$$rate = k[A][B]^3$$

This reaction is first order in A, third order in B and fourth order overall $(1 + 3 = 4)$. Thus, the reaction order represents how the reaction depends on the reactant concentrations.

- Zero-order reactions

Reactions whose order is zero are rare in heterogeneous catalysis and biocatalysis. In a zero-order reaction,

$$A \rightarrow products$$

The reaction rate is

$$rate = -\frac{\Delta[A]}{\Delta t}$$

The rate law is given by

$$rate = k[A]^0 = k$$

Thus, the rate of a zero-order reaction is a constant independent of the reactant concentration. Using calculus from the above equations,

$$-k\int_0^t dt = \int_{[A]_0}^{[A]_t} d[A]$$

or in the linearized form,

$[A]_t = -kt + [A]_0$.

As shown in Fig. 3.2, a plot of $[A]_t$ versus time gives a straight line with slope $= -k$ and y-intercept $= [A]_0$.

Another evaluation of the reaction rate relating the concentration to time is the half-life $(t_{1/2})$, which is defined as the time required for the concentration of a reactant to decrease to half of its initial concentration. The concept of a half-life as applied to catalysts is the time when the catalyst activity has been reduced to one-half. Thus, the half-life of a zero-order reaction can be written as follows:

Fig. 3.2 Zero-order reaction: plot of $[A]_t$ versus time. The slope of the line is equal to $-k$. Adapted from Piumetti and Russo [3]

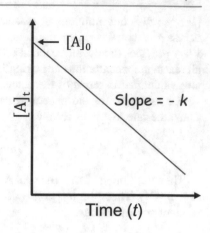

$$t_{1/2} = \frac{[A]_0}{2k}$$

- First-order reactions

A first-order reaction is a reaction whose rate depends on the reactant concentration raised to the first power. In a first-order reaction,

$$A \rightarrow products$$

The reaction rate is

$$rate = -\frac{\Delta[A]}{\Delta t}$$

The rate law is

$$rate = k[A]$$

By combining the last two equations,

$$k[A] = -\frac{\Delta[A]}{\Delta t}$$

In differential form, the equation becomes

$$k[A] = -\frac{d[A]}{dt}$$

By rearranging the above equation,

$$-kdt = \frac{d[A]}{[A]}$$

Integrating between $t = 0$ and $t = t$ gives

$$-k \int_0^t dt = \int_{[A]_0}^{[A]_t} \frac{d[A]}{[A]}$$

$$\ln \frac{[A]_t}{[A]_0} = -kt$$

or

$$\ln[A]_t - \ln[A]_0 = -kt$$

Thus, the equation can be written in the linearized form

$$\ln[A]_t = -kt + \ln[A]_0$$

If $\ln[A]_t$ is plotted as a function of time, a straight line is obtained with a slope equal to $-k$ and a y-intercept equal to $\ln[A]_0$ as shown in Fig. 3.3. The rate constant is obtained from the slope of this plot.

According to the definition of a catalyst half-life, when $t = t_{1/2}$, $[A]_t = [A]_0/2$; hence, for a first-order reaction, $t_{1/2}$ is

$$t_{1/2} = \frac{1}{k} \ln \frac{[A]_0}{[A]_0/2}$$

Fig. 3.3 First-order reaction: plot of $\ln[A]_t$ versus time. The slope of the line is equal to $-k$. Adapted from Piumetti and Russo [3]

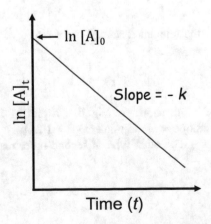

that is,

$$t_{1/2} = \frac{1}{k}\ln 2 = \frac{0.693}{k}$$

• Second-order reactions

In a second-order reaction, the rate depends on the concentration of one reactant raised to the second power or the concentrations of two reactants each raised to the first power. The simpler second-order reaction involves only one reactant as follows:

$$A \rightarrow products$$

in which

$$rate = -\frac{\Delta[A]}{\Delta t}$$

From the rate law,

$$rate = k[A]^2$$

In differential form, the equation becomes

$$k[A]^2 = -\frac{d[A]}{dt}$$

By integrating between $t = 0$ and $t = t$,

$$-k\int_0^t dt = \int_{[A]_0}^{[A]_t} \frac{d[A]}{[A]^2}$$

the integrated second-order rate equation is

$$\frac{1}{[A]_t} = kt + \frac{1}{[A]_0}$$

As shown in Fig. 3.4, a plot of $1/[A]_t$ versus time gives a straight line with slope $= k$ and y-intercept $= 1/[A]_0$.

The other type of second-order reaction involves two reactants as follows:

$$A + B \rightarrow products$$

Fig. 3.4 Second-order reaction for A → products: plot of $1/\ln[A]_t$ versus time. The slope of the line is equal to k. Adapted from Piumetti and Russo [3]

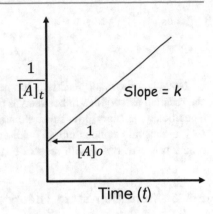

The rate law is

$$rate = k[A][B]$$

This reaction is first order in A and first order in B for an overall reaction of order 2. However, a second-order reaction rate with reactants A and B can be problematic; the concentrations of the two reactants must be evaluated simultaneously, which may be difficult. A common solution to this problem is the pseudo-first-order approximation.

One way to obtain a pseudo-first-order reaction is to use a large excess of one reactant such that as the reaction occurs, only a small amount of the excess reactant is consumed, and its concentration can be considered constant. For example, if $[B] \gg [A]$, $[B]$ will be effectively constant such that

$$rate = k[A][B] = k'[A]$$

Thus, the rate law has the appearance of a first-order reaction. The rate constant (k') is given by $k[B]$, where k is the second-order rate constant. If k' is measured at many different initial concentrations of B, a plot of k' versus $[B]$ yields a straight line with a slope equal to k. For example, the hydrolysis of esters by dilute mineral acids occurs via pseudo-first-order kinetics, where the concentration of water is present in large excess.

$$CH_3COOCH_3 + H_2O \rightarrow CH_3COOH + CH_3OH$$

This is the case of hydrolysis reactions catalysed by enzymes in aqueous media.

The half-life of a second-order reaction of type A → B can be written as follows:

$$\frac{1}{[A]_o/2} = kt_{1/2} + \frac{1}{[A]_0}$$

Thus,

$$t_{1/2} = \frac{1}{k[A]_0}$$

Notably, enzymatic reactions are neither zero-order nor first-order; the order of the reaction usually lies in between as described by the Michaelis–Menten equation. Specifically, at (very) low substrate amounts, the reaction is first order with respect to the substrate concentration, while at (very) high substrate concentrations, the reaction is zero-order with respect to the substrate as presented in Chap. 2.

3.2.2 Transition State Theory [9]

In order to react, the colliding molecules must have a total kinetic energy that is equal to or greater than the activation energy (E_a), which is defined as the amount of energy needed to initiate a chemical reaction. Therefore, the activation energy is a barrier that must be overcome for a reaction to occur. The presence of a catalyst allows a decrease in the activation energy, thereby increasing the reaction rate. Therefore, catalysts act as a "doping drug" in chemical reactions, favouring the transformation of reactants into products.

When reactant molecules collide, they create an activated complex with a range of configurations before the product is formed. The activated complex of a reaction may refer to either the transition state, i.e., the state corresponding to the highest potential energy along the reaction coordinate (usually denoted as $\ddagger$), or to other states along the reaction coordinate between reactants and products; in particular those close to the transition state.

The transition state refers to the highest potential energy configuration of the atoms during the reaction only, whereas the activated complex represents a range of configurations near the transition state which the atoms pass through during the transformation from reactants into products. This may be described in terms of a reaction coordinate, in which the transition state is the molecular configuration at the peak of the diagram, and the activated complex represents any point near the maximum (Fig. 3.5).

The activated complex to which the structure $[AB]^{\ddagger}$ is assigned for the reaction

$$A + B \rightarrow [AB]^{\ddagger} \rightarrow C + D$$

has a higher energy (enthalpy) than the reactants. If the products are more stable than the reactants, the reaction is exothermic ($\Sigma H_{products} - \Sigma H_{reactants} < 0$). In contrast, if the products are less stable than the reactants, the process is endothermic ($\Sigma H_{products} - \Sigma H_{reactants} > 0$).

The transition states of chemical reactions seem to have lifetimes near 10^{-13} s, which is on the order of magnitude of the time of a single bond vibration. It appears that enzymes act to stabilize transition states that lie between reactants and products

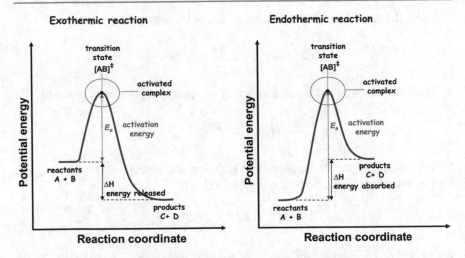

Fig. 3.5 Potential energy diagrams of exothermic and endothermic reactions

and, thus, are expected to bind any inhibitor closely resembling such a transition state. Substrates and products often engage in several enzyme reactions, while the transition state has a tendency to be a characteristic of one specific enzyme.

The analysis of a unimolecular reaction is presented here. Suppose that $\Delta G^{\ddagger}$ is the difference in Gibbs free energy between the transition state $X^{\ddagger}$ and the ground state X (namely, the Gibbs free energy of activation). The relationship between the Gibbs free energy change and the equilibrium constant is

$$\Delta G = -RT \ln K$$

Then, from equilibrium thermodynamics,

$$[X^{\ddagger}] = [X] e^{-\Delta G^{\ddagger}/RT}$$

The transition state decomposes at the vibrational frequency (v) of the bond that is breaking. This frequency can be derived from the equivalence of the energies of an excited oscillator which can be calculated from quantum theory and classical physics as follows:

$$E = hv = kT$$

where k and h are the Boltzmann and Plank constants, respectively.

Therefore,

$$v = \frac{kT}{h}$$

At 25 °C $v \sim 10^{12} \ s^{-1}$.

Then, the rate of the decomposition of X can be calculated by

$$-\frac{d[X]}{dt} = v[X^{\ddagger}] = \frac{kT}{h}[X]e^{-\Delta G^{\ddagger}/RT}$$

Therefore, the first-order rate constant k_1 of the decomposition is given by

$$k_1 = \frac{kT}{h}e^{-\Delta G^{\ddagger}/RT}$$

Moreover, $\Delta G^{\ddagger}$ can be separated into enthalpic and entropic terms by using the equilibrium thermodynamic relationship as follows:

$$\Delta G^{\ddagger} = \Delta H^{\ddagger} - T\Delta S^{\ddagger}$$

where $\Delta H^{\ddagger}$ and $\Delta S^{\ddagger}$ are the differences in enthalpy and entropy, respectively. Then, the rate constant becomes

$$k_1 = \frac{kT}{h}e^{\Delta S^{\ddagger}/R}e^{-\Delta H^{\ddagger}/RT}$$

A more rigorous approach must include the transmission coefficient ($\gamma \sim 1$ for simple reactions), i.e., a multiplication factor corresponding to the fraction of the transition state that proceeds into the products [9].

3.2.3 The Arrhenius Equation

The dependence of the reaction rate constant on temperature can be expressed by the Arrhenius equation as follows:

$$k = Ae^{-E_a/RT}$$

where E_a is the activation energy of the reaction (in kJ/mol), R is the universal gas constant (8.314 J/K·mol), and T is the absolute temperature (K). The pre-exponential factor A which represents the collision frequency, is called the frequency factor, and depends on how often the molecules collide when all concentrations are 1 mol L^{-1} and whether the molecules are properly oriented when they collide. If the reaction is first order, it has units of s^{-1}. A can be considered constant in a given reaction system over a fairly wide temperature range.

The Arrhenius equation can be expressed in a more useful form by applying the natural logarithm to both sides as follows:

$$\ln k = \ln A - \frac{E_a}{RT}$$

Fig. 3.6 Arrhenius plot of
the logarithm of the rate
constant versus the inverse of
temperature for a given
reaction. Adapted from
Piumetti and Russo [3]

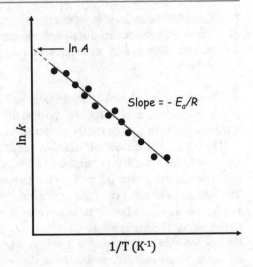

Thus, a plot of $\ln k$ versus $1/T$ gives a straight line whose slope is equal to $-E_a/R$ and y-intercept is $\ln A$ (Fig. 3.6). Consequently, the activation energy can be determined from the slope of the plot.

3.3 Fundamental Concepts in Heterogeneous Catalysis [5, 9–11]

Catalysis originated as a scientific discipline in the early part of the last century, which has been described as "the age of the molecularization of the sciences". However, almost a century passed before the molecular basis of some catalytic processes that are currently widely applied at a large scale were understood. Due to recent developments in characterization techniques and computational methods, great progress has been achieved in the research and development of catalytic processes, and new concepts have been introduced in molecular catalysis.

3.3.1 Steps in Heterogeneous Catalysis

During a catalytic reaction, the reactants and products undergo a series of steps over the catalyst surface [3], including the following:

1. External diffusion of molecules (across the boundary layer surrounding the catalyst).
2. Diffusion of molecules through the internal porosity of the catalyst.
3. Adsorption of molecules onto active sites.
4. Catalytic reactions at active sites.

5. Desorption of molecules from active sites.
6. Diffusion of molecules through the internal cavities of the catalyst.
7. External diffusion of molecules (across the boundary layer surrounding the catalyst).

These steps are shown schematically in Fig. 3.7.

Moreover, chemical reactions are usually accompanied by heat effects, which typically occur in porous catalysts and across the external boundary layer.

The kinetics of the overall reaction (steps 1–7) are usually called macrokinetics (or sometimes effective or apparent kinetics), while the steps of the kinetics of catalytic transformation (steps 3–5) are termed microkinetics (or intrinsic kinetics). The intrinsic chemical rates are exponential functions of temperature according to the Arrhenius law, whereas mass transfer phenomena are less strongly affected by temperature. The intraparticle diffusivity D_e is proportional to $T^{1.5}$ when molecular diffusion dominates, whereas it is proportional to $T^{0.5}$ when governed by Knudsen diffusion, which involves molecular collisions with the pore wall rather than among molecules.

Figure 3.8 shows an Arrhenius diagram and a typical conversion trend as a function of temperature in a catalytic fluid–solid reaction conducted using a porous catalyst.

At a low temperature, the catalytic reaction is slow and rate controlling. In this region of kinetic control, the slope is equal to $-E_a/R$. As the temperature increases, the intrinsic chemical kinetics are accelerated more strongly than intraparticle (or internal) diffusion.

As a result, the measured (apparent) activation energy in this regime is approximately half the true (effective) value (slope $= -E_a/2R$). In this internal diffusion regime, it is often advantageous to use microporous or mesoporous catalysts with a 3-D framework in which the active sites are highly accessible. Finally, at a high temperature, interphase mass transfer becomes the rate-controlling step. The slope of the curve is approximately zero, corresponding to an apparent activation energy in the range of less than 5–10 kJ mol^{-1}.

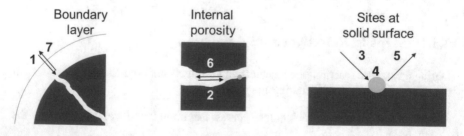

Fig. 3.7 Fundamental steps of a catalytic fluid–solid reaction performed in a porous catalyst. Adapted from Piumetti and Russo [3]

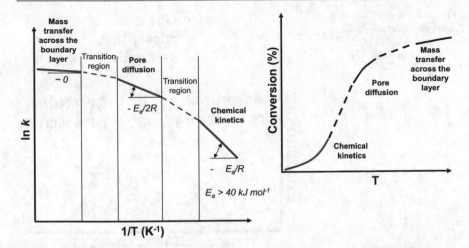

Fig. 3.8 Transition from the kinetic-controlled regime to the diffusion-controlled regime in a heterogeneous catalytic reaction. Arrhenius plot (left); conversion trend versus temperature (right). Adapted from Piumetti and Russo [3]

3.3.2 The Sabatier Principle

The Sabatier principle states that an effective catalyst should exhibit an interaction of intermediate strength with the reactants, products, and intermediates of the catalytic reaction (Fig. 3.9). According to this principle, the interactions between the catalyst surface and the various adsorbed species should be "just right", i.e., neither too strong nor too weak. If the interaction is too weak, the surface reactions have high activation energies, resulting in low catalytic activity. In contrast, if the interaction is too strong, excessive blocking of the catalyst surface sites occurs, similarly leading to low catalytic activity. Thus, the catalytic rate reaches a maximum at the optimum adsorption strength. The optimal set of surface binding energies depends on the selected set of reaction conditions.

For example, plotting the rate of decomposition of formic acid on several metals versus the heat of the formation of metallic formates (ΔH_f) yields a volcano-shaped curve (Fig. 3.10). At low ΔH_f values, the reaction rate is low and corresponds to the rate of adsorption, which increases with the increasing heat of the formation of formates (representing the stability of the surface compound). At high ΔH_f values, the reaction rate is also low and corresponds to the desorption rate, which increases with decreasing ΔH_f.

3.4 Chemisorption and Physisorption

Chemisorption consists of a chemical interaction between a gas-phase molecule and a solid surface. This phenomenon typically occurs in heterogeneous catalysts. When new chemical bonds are formed at the solid surface, the adsorbate is referred

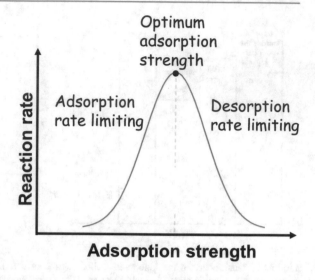

Fig. 3.9 Reaction rate as a function of the adsorption strength. Adapted from Piumetti and Russo [3]

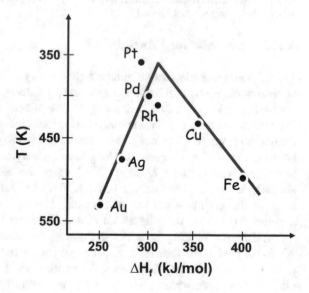

Fig. 3.10 Volcano plot of the decomposition of formic acid on several metals. The temperature at which the reaction reaches a specific rate was used as a comparative value. Drawn based on data extracted from [12]. Adapted from Piumetti and Russo [3]

to as 'chemisorbed'. The chemisorption energy is approximately 30–70 kJ mol^{-1} for molecules and 100–400 kJ mol^{-1} for atoms. However, when chemical bond formation is weak or no chemical bond is formed, the process is referred to as "physisorption". Molecules interact with solid surfaces through van der Waals forces, and the interaction energy is typically 5–10 kJ mol^{-1}. In contrast to chemisorbed species, physisorbed species (physisorbates) do not exhibit directional interactions. Table 3.3 summarizes the main aspects of chemisorption and physisorption phenomena.

Table 3.3 Comparison of chemisorption and physisorption

Properties	Chemisorption	Physisorption
Nature of adsorption	– Chemical bond formation – Electron exchange – Strong interaction – Often dissociative – Irreversible	– van der Waals forces – Polarization – Weak interaction – Not dissociative – Reversible
Interaction energy	– 30–70 kJ mol^{-1} (for molecules) – 100 – 400 kJ mol^{-1} (for atoms)	< 30 kJ mol^{-1}
Adsorption temperature	Unlimited range	$\leq T_{bp}$ of adsorbate
Crystallographic specificity	Great differences among crystal planes (high directionality)	Independent of surface geometry (no directionality)

Adapted from Piumetti and Russo [3]
[a] T_{bp} is the boiling temperature of the adsorbate

The energetics of dissociative adsorption can be illustrated by the one-dimensional potential plot proposed by Lennard–Jones as shown in Fig. 3.11. When A_2 (diatomic molecule) reaches a catalytic surface, it first experiences weak bonding as $A_{2,ad}$. The dissociation of the A_2 molecule requires dissociation energy (E_{diss}), and then, 2A atoms can form strong interactions with the catalytic surface (A_{ad}).

3.4.1 Langmuir–Hinshelwood Versus Eley–Rideal Mechanism [11]

The Langmuir–Hinshelwood (LH) mechanism is used to describe numerous surface processes in which reactants (e.g., hydrocarbon and oxygen) interact after adsorption onto uniform surfaces with one or more types of sites; in contrast, in the Eley–Rideal (ER) mechanism, adsorbed reactant A (chemisorbed species) interacts directly by collision with reactant B impinging from the gas phase (Fig. 3.12).

In homogeneous catalysis, the reagent concentrations are usually well known, while in heterogeneous catalysis, the concentrations of the reactants correspond to the adsorbed phases rather than to those in the reaction medium. According to Henry's law, it can be assumed that this concentration is proportional to the partial pressure of the reactants (p_i) on the catalyst surface.

Considering the reaction $A + B \rightarrow C$ occurring at a solid surface, the reaction rates according to either the LH- or ER-type mechanisms are as follows:

$$r = k\theta_A\theta_B \text{ (Langmuir-Hinshelwood)}$$

and

$$r = k'\theta_A p_B \text{ (Eley-Rideal)}$$

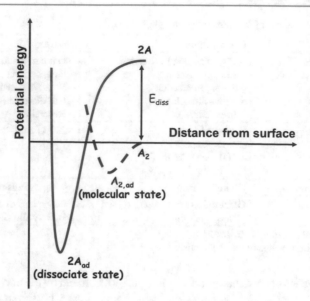

Fig. 3.11 Lennard–Jones's model diagram ilustrating the energetics of dissociative adsorption on a surface. Adapted from Ertl [10]

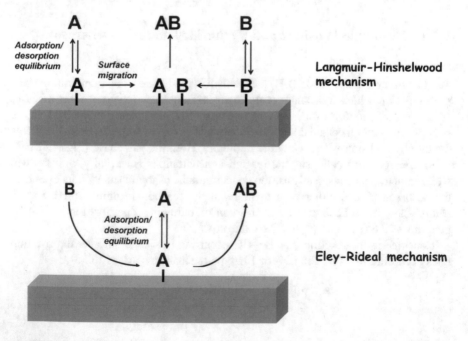

Fig. 3.12 Scheme of the Langmuir–Hinshelwood and Eley–Rideal mechanisms. Adapted from Piumetti and Russo [3]

where the coverages θ_A and θ_B are related to the respective partial pressures p_A and p_B as follows:

$$\theta_A = \frac{b_A p_A}{1 + b_A p_A + b_B p_B}$$

$$\theta_B = \frac{b_B p_B}{1 + b_A p_A + b_B p_B}$$

Therefore, an adsorption site may be occupied by either A or B, and the product molecule is so weakly adsorbed that its surface concentration is negligible. Thus, the LH equation becomes

$$r = k\theta_A \theta_B = \frac{b_A b_B p_A p_B}{(1 + b_A p_A + b_B p_B)^2}$$

For fixed values of T and p_B, the reaction rate passes through a maximum as p_A is varied. The maximum occurs when A and B have equal surface concentrations, i.e., $\theta_A = \theta_B$.

However, with an ER-type mechanism,

$$r = k'\theta_A p_B = \frac{k' b_B p_A p_B}{1 + b_B p_B}$$

which shows that the rate is first-order with respect to p_A and that the order with respect to p_B changes from first order to zero order.

In catalytic reactions of type $A_{(ad)} \rightarrow B + C$, which involve the breaking of chemical bonds, the reaction rate becomes

$$r = \frac{k b_A p_A}{1 + b_A p_A}$$

This equation is comparable to the Michaelis–Menten model proposed to describe the kinetics of enzyme catalysis as a function of the substrate concentration (*vide* Chap. 2).

Two limiting cases may occur. If $b_A p_A \ll 1$, $r = k b_A p_A$, and the reaction is kinetically first order (very low surface coverage); if $b_A p_A \gg 1$, $r = k$, and the reaction is zero order in gas pressure (almost complete surface coverage) (Fig. 3.13).

Most surface catalytic reactions occur via the LH mechanism in which the rate is controlled by the reaction of the adsorbed molecules when the adsorption and desorption pressures are in equilibrium. For example, the reaction $A_2 + 2B \leftrightarrow 2AB$ may occur via the following LH-type mechanism:

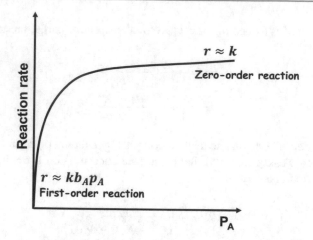

Fig. 3.13 Langmuir-Hinshelwood type model of the reaction $A_{(ad)} \rightarrow B + C$: two limiting cases. Adapted from Piumetti and Russo [3]

$$A_2 + * \leftrightarrow A_2*$$
$$A_2 * + * \leftrightarrow 2A*$$
$$B + * \leftrightarrow B*$$
$$A * + B* \leftrightarrow AB * + *$$
$$AB* \leftrightarrow AB + *$$

where the asterisks denote the active centres. The same reaction may also occur via an ER-type mechanism as follows:

$$A_2 + * \leftrightarrow A_2*$$
$$A_2 * + * \leftrightarrow 2A*$$
$$A * + B \leftrightarrow AB + *$$

where the last step is the direct reaction between the adsorbed species (A*) and molecule B (in the gas phase).

If the reaction rate is measured as a function of coverage by A*, it initially increases according to both mechanisms (Fig. 3.14). In the ER mechanism, the rate increases with increasing coverage until the surface is completely covered by species A, while in the LH mechanism, the rate reaches a maximum and decreases to zero when the surface is completely covered by species A because step B + * B* cannot occur when A* blocks all active centres. Thus, it is possible to determine the "true" kinetic mechanism of catalytic reactions.

The LH mechanism has a counterpart in enzyme kinetics, i.e., the so-called ternary-complex mechanism (*vide* Chap. 2), where two substrate molecules (species

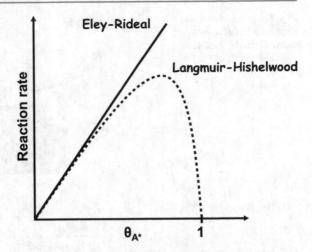

Fig. 3.14 Study of reaction mechanisms according to the Eley–Rideal and Langmuir–Hinshelwood models. Adapted from Piumetti and Russo [3]

A and B) bind the surface of one giant molecule (enzyme E) to yield the species EAB, which evolves to EYZ and then eventually decomposes into the enzyme and products Y and Z.

3.4.2 Mars and Van Krevelen Mechanism [11, 13, 14]

The Mars and van Krevelen (MvK) mechanism (or redox-type mechanism) is widely applied in heterogeneous oxidation catalysis. A hydrocarbon molecule (R-H) reacts by extracting lattice oxygen (O^{2-}) from the surface (R-H + O^{2-} → R-O$^-$ + H$^+$ + 2e$^-$), thereby generating oxidized products and a reduced catalyst surface. Then, the lattice oxygen is replenished by the reduction of gaseous oxygen (½ O_2 + 2e$^-$ → O^{2-}) (Fig. 3.15).

The substrate is oxidized by a solid catalyst (e.g., metal oxide catalysts) and not directly by molecular oxygen in the gaseous phase.

The mechanism can be expressed as follows:

$$R + \beta O_L \rightarrow P + \beta \ \blacksquare \ r_{red} = k_{red} p_R \theta$$

$$\blacksquare + \frac{1}{2} O_2 \rightarrow O_L r_{ox} = k_{ox} p_{O_2}^n (1 - \theta)$$

where β is the stoichiometric requirement for lattice oxygen to generate the product P; O_L and $\blacksquare$ represent the lattice oxygen anion and the corresponding vacancy, respectively; and R is the reactant to be oxidized.

By introducing the steady state condition $r_{red} = r_{ox}/\beta$, we obtain the following:

$$\beta k_{red} p_R \theta = k_{ox} p_{O_2}^{\ n} (1 - \theta)$$

Fig. 3.15 Scheme of the Mars and van Krevelen mechanism. Adapted from Piumetti and Russo [3]

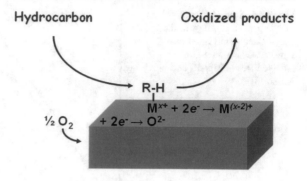

Therefore,

$$\theta = \frac{k_{ox}p_{O_2}{}^n}{\beta k_{red}p_R + k_{ox}p_{O_2}{}^n}$$

Thus, the final equation obtained is

$$\frac{1}{r_{red}} = \frac{1}{k_{red}p_R} + \frac{\beta}{k_{ox}p_{O_2}{}^n}$$

and is commonly used to describe the rate equation of the MvK mechanism, which is widely used in oxidation catalysis.

The MvK mechanism occurs frequently over VO_x-based catalysts. However, in addition to oxidation, reduction, deoxygenation and oxidative dehydrogenation reactions, many other catalytic processes, including chlorination, dechlorination and hydrodesulfurization, can be described by MvK kinetics [15].

Examples of the MvK-like mechanism are also known in homogeneous oxidation catalysis, e.g., in Wacker chemistry, typically consisting of the conversion of ethylene into acetaldehyde by oxygen in water in the presence of Pd^{2+} species. Moreover, the MvK-type mechanism has a counterpart in enzyme kinetics, i.e., the so-called "ping-pong" mechanism (see Chap. 2), because the enzyme species E reacts with substrate A, which yields Y, turns into the species E* (enzyme intermediate), and is restored to E by reaction with substrate B, which turns into Z [15].

3.4.3   Deactivation of Catalytic Action [3]

Catalyst deactivation is a complex phenomenon that may include not only activity loss but also decreased selectivity. Heterogeneous catalysts are usually deactivated during time-on-stream (TOS) reactions, and the time scale of deactivation strongly depends on the type of catalytic process. Overall, there are several causes for catalyst deactivation, including *inhibition, poisoning, formation of deposits, thermal degradation, mechanical degradation, and corrosion/leaching* by the reaction mixture.

The abovementioned causes are often interdependent and may result in either reversible catalyst deactivation (e.g., catalyst fouling) or irreversible deactivation (e.g., mechanical deactivation). Leaching reflects catalyst loss, which leads to catalytic inactivation. However, deposits (e.g., dust) and sintering phenomena do not affect the rate constant but reduce both the accessible active sites and the catalyst effectiveness.

Overall, an *inhibitor* is a substance that reduces the rate of a catalytic reaction, typically as a consequence of bonding chemically to the catalyst. This alteration, reduction or suppression of catalytic action can be due to the adsorption of inhibitory or deactivating substances on active sites along with their possible modification and restriction of physical access to active sites due to the presence of inhibitors. Notably, a characteristic feature of inhibition is its reversible nature. Competitive inhibitors that slow down reactions by competing with reactants in bonding to catalysts are also known. However, a *catalyst poison* is usually defined as a substance that interacts strongly with a catalyst. Indeed, a distinct feature of poisoning is its irreversible nature. Thus, poisoning is deactivation caused by the chemisorption of substances (i.e., impurities) on the catalyst surface. Additionally, reactants and products can be strongly adsorbed on active sites and may hinder the adsorption of weaker adsorbing species.

Thus, inhibition and poisoning are differently defined in heterogeneous catalysis, whereas in biocatalysis, inhibitors include substances that interact either weakly or strongly with enzymes (vide Chap. 2). Thus, both reversible and irreversible inhibitors may interact with enzymes, whereas reversible inhibition is typically observed in heterogeneous catalysts.

Moreover, it is possible to distinguish between nonselective and selective poisoning phenomena. Nonselective poisoning of the catalyst surface leads to a decrease in the total number of active sites (N_T) as follows:

$$N = N_T(1 - \alpha)$$

where α is the fraction of sites that are poisoned. Therefore, the activity decreases linearly with the fraction of poisoned sites. In contrast, selective poisoning may occur when there is preferential adsorption of the poison on the most active centres.

The *formation of deposits* is another important cause of catalyst deactivation and depends on the operating conditions. Fouling is a process in which substances (e.g., carbonaceous deposits) are deposited on the catalyst surface, thereby blocking the active sites and/or pores. Coke is usually formed by chemical reactions with hydrocarbons at relatively high temperatures. These reactions leave a layer of hydrogen-deficient carbonaceous compounds on the catalyst surface, rendering the active sites inaccessible to reactants. In addition to coke, other deposits, including carbonates, sulfates and nitrates, can be found on the catalyst surface.

Thermal degradation is a phenomenon leading to catalyst deactivation as a consequence of high-temperature-induced sintering, chemical changes, and evaporation. Sintering is strongly temperature-dependent and results in a loss of surface area due to crystallite growth of the catalyst. Sintering may occur in both supported and unsupported catalytic materials.

Mechanical degradation is often observed in solid catalysts. In fact, catalytic materials in reactors are mechanically stressed over their life cycle (i.e., during start-up, cooling and regeneration steps).

In liquid-phase catalytic reactions, the *leaching* of active components into the reaction medium by dissolution is often the main concern. Leaching is the result of the solvolysis of the metal–oxygen bonds through which the catalyst is anchored to the solid support. The support may also be subject to leaching. The reaction medium can also be corrosive. For example, alumina at high or low pH may dissolve, leading to corrosion and leaching effects.

3.5 Summary

- Nanocatalysis, which is also known as "semi-heterogeneous" catalysis, represents a bridge between homogeneous and heterogeneous catalysis.
- Enzymes have features of both homogenous and heterogeneous catalysts. Similar to heterogeneous catalysts, enzymes have an active surface on which reactants are temporarily adsorbed while waiting for the subsequent steps. Similar to homogenous catalysts, enzymes' amino acid groups actively interact with substrates in multistep sequences.
- A catalytic reaction can be described by modelling the macroscopic kinetics by correlating atomic processes with macroscopic parameters (e.g., partial pressures and temperature) within the framework of a continuum model. The kinetic model is based on an accurate description of the catalytic cycle. Enzymatic reactions are neither zero- nor first-order reactions but usually in between as described by the Michaelis–Menten equation.
- The Langmuir–Hinshelwood mechanism is used to describe numerous surface processes in which reactants interact after adsorption onto uniform surfaces with one or more types of sites. The counterpart of this mechanism in enzyme kinetics is the so-called ternary-complex mechanism.
- The Mars and van Krevelen mechanism is widely applied in heterogeneous oxidation catalysis and homogeneous oxidation catalysis, e.g., in Wacker chemistry. The counterpart of this mechanism in enzyme kinetics is the so-called "ping-pong" mechanism.

3.6 Questions

1. Table 3.1 reports the main advantages and disadvantages of homogeneous catalysis. Similarly, can you visualize the main advantages and disadvantages of immobilized enzymes over free enzymes in biotech processes?
2. Why does nanocatalysis represent an important bridge between homogeneous and heterogeneous catalysis?

3. Enzymes have features of both homogenous and heterogeneous catalysts. What does this mean in your opinion?
4. The counterpart of the Langmuir–Hinshelwood mechanism in enzyme kinetics is the so-called ternary-complex mechanism. Summarize the main features of the two kinetic mechanisms including their similarities and differences.

References

1. G. Ertl, H. Knözinger, F. Schütz, J. Weitkamp, *Handbook of Heterogeneous Catalysis*, 2nd edn. (Wiley-VCH, 2008)
2. J.M. Thomas, W.J. Thomas, *Principles and Practice of Heterogeneous Catalysis*, 2nd edn. (Wiley-VCH, 2014)
3. M. Piumetti, N. Russo, *Notes on Catalysis for Environment and Energy* (CLUT – Politecnico, 2018)
4. G.F. Swiegers, *Mechanical Catalysis: Methods of Enzymatic, Homogeneous and Heterogeneous Catalysis* (Wiley-VCH, 2008)
5. J.C. Védrine, Appl. Catal. A **474**, 40–50 (2014)
6. R.A. van Santen, M. Neurock, *Molecular Heterogeneous Catalysis* (Wiley-VCH, 2006)
7. M.S. Silberberg, *Chemistry*, 5th edn. (McGraw-Hill, 2009), pp. 686–736
8. J.K. Nørskov, F. Studt, F. Abild-Pedersen, T. Bligaard, *Fundamental Concepts in Heterogeneous Catalysis* (Wiley-VCH, 2014)
9. A. Fersht A, *Enzyme Structure and Mechanism* (W.H. Freeman, San Francisco, 1985), pp. 54–57
10. G. Ertl. *Reactions at Solid Surfaces* (Wiley-VCH, 2009)
11. M. Beller, A. Renken, R. van Santen, *Catalysis: From Principles to Applications* (Wiley-VCH, 2012)
12. H. Knözinger, K. Kochloefl, Heterogeneous catalysis and solid catalysts, in *Ullmann's Encyclopedia of Industrial Chemistry* (Wiley-VCH, 2005)
13. B.K. Hodnett, *Heterogeneous Catalytic Oxidation* (Wiley-VCH, 2000)
14. P. Mars, D.W. van Krevelen, Chem. Eng. Sci. Spec. Suppl. **3**, 41–59 (1954)
15. M. Piumetti, F.S. Freyria, B. Bonelli, Chimica Oggi—Chem. Today **31**, 55–58 (2013)

Complex Nature of Active Sites

4

4.1 Dynamic Behaviour of Active Sites [1, 2]

Heterogeneous catalysts behave as dynamic systems that adapt their structure to specific operating conditions and, therefore, are smart materials.

Over the last few decades, due to the development of modern scanning probe techniques and advanced methods for in situ characterization, it has become possible to study the dynamic behaviour of active sites and observe catalytic surfaces on the atomic scale. For example, scanning tunnelling microscopy (STM) has allowed the visualization of steps, such as dissociation at "active sites" and the surface diffusion of adsorbates, at the atomic level. Similarly, more recently, 4-D electron microscopy, which combines an ultrafast laser with electron microscopy, has allowed investigations of the behaviour of single atoms at the femtosecond time scale. As a result, these recent advanced techniques have reinforced the awareness that active sites are dynamic entities. However, except for the simplest gas-phase reactions, detailed knowledge of reaction mechanisms is still very difficult to achieve. During a reaction, indeed, the catalyst surface usually interacts with a mix of fluctuating molecules (e.g., reactants, products, by-products, and intermediates). Moreover, the more complex the chemical reaction is, the slower the TOF as the reaction steps involve several molecular rearrangements. In fact, the catalyst surface must be regenerated after each reaction cycle through product desorption, followed by the reorganization of active sites and their environment [1, 2].

Electronic Supplementary Material The online version of this chapter (https://doi.org/10.1007/978-3-030-88500-7_4) contains supplementary material, which is available to authorized users.

4.2 Active Sites in Heterogeneous Catalysis: Historical Background [1, 2]

According to the Langmuir model, the catalyst surface can be viewed as an array of energetically identical sites that do not interact with each other. Langmuir imagined the surface as a "checkerboard", with the ideal assumption that solid surfaces exhibit identical and noninteracting sites.

A remarkable step forward occurred in 1925 when H. S. Taylor introduced the concept of active sites while proposing the existence of surface atoms with different degrees of saturation by neighbouring atoms. In contrast to the Langmuirian view of a uniform distribution of static sites that do not interact with one another, Taylor's theory proposed vacancies and surface atom configurations as centres of reactivity.

As a result, it was established that the exposed faces of solid catalysts typically contain face (or terrace), edge (or step) and corner (or kink) sites that have different coordination numbers (CNs). The CN may also change, leading to different reactivities to the corresponding sites.

Subsequently, it became clear that preferential adsorption occurs at the step and kink sites of solid surfaces rather than at terraces (Fig. 4.1). However, the sites capable of developing the strongest interactions with reactants are not always the most active; in fact, an intermediate binding energy allows the optimization of the residence time of the adsorbate and promotes its conversion into products [3].

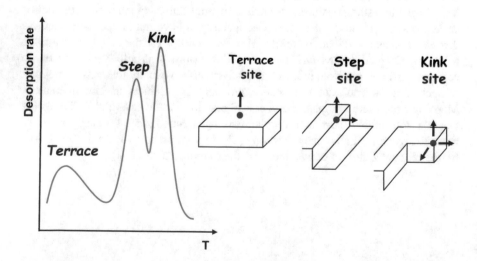

Fig. 4.1 A temperature-programmed desorption pattern showing three main peaks corresponding to progressively stronger bound states of an adsorbed organic molecule at the terrace, step and kink sites on a roughened solid surface. Adapted from Thomas et al. [3]

Subsequently, G. A. Somorjai showed that the step and kink sites of surface metals can be very active in breaking C–H and C–C bonds in hydrocarbon molecules. Furthermore, M. Boudart et al. introduced the concepts of structure-sensitive and structure-insensitive reactions: the lack of an effect of the particle size was first observed in the hydrogenation of cyclopropane on single crystals of platinum, whereas the role of the dimensions in the synthesis of ammonia over iron-based systems was confirmed.

It has since been observed that C–C or N–N bonds breaking typically involves structure sensitivity, and the active sites consist of ensembles of surface atoms, whereas the breaking or formation of H–H, C–H or H–O bonds is structure insensitive, and the active sites are simpler, usually isolated atoms [5].

The pioneering studies by O. Beeck et al. showed a correlation between the catalytic activity for ethene hydrogenation and the d-character of several transition metals. In particular, Beeck argued that catalytic activity is related to the presence of holes in the d-band and the geometry of the catalyst surface. Consequently, many studies were conducted to clarify the electronic aspects of solid metals, and two principal models were proposed to describe their behaviour, namely, (i) the "atomistic model" and (ii) the "band model". The atomistic model does not consider the entire solid but focuses on surface sites consisting in a single atom or a small group of atoms (cluster), while the band model describes the solid surface in terms of surface states and the localized energy levels available at the surface without considering the interactions between individual atoms.

Geometric and electronic properties are critical for catalytic activity. These two properties are often considered separately, although their interrelation in many catalytic reactions (e.g., selective oxidations and isomerizations) has been well established.

It is also known that an active site may include more than a single catalytic centre as it may consist of clusters of atoms (e.g., metals, metal ions, anions, Lewis/Brønsted sites, acid–base pairs, or organometallic compounds). For instance, in the case of H_2 dissociation on a Pd(111) surface, an aggregate of three or more vacant sites is needed to form an active site. Similarly, sulfonic acid resins presenting a dense network of $-SO_3H$ groups typically exhibit higher catalytic activity for reactions catalysed by hydrated protons than resins with isolated $-SO_3H$ groups, highlighting the important role of a H-bonded network [1]. Another example was described by Grasselli et al. [4] who applied a vanadyl pyrophosphate (VPO) catalyst for the oxidation of butane to maleic anhydride. This catalytic reaction involves a 14-electron change, 8 H abstractions and 3 O insertions, and it appears that the active sites for this selective oxidation consist of four VPO dimer sites separated from the other sites by excess surface P entities [5]. As shown in Fig. 4.2, there are favourable pathways for surface oxygen mobility within the sites, whereas P_2O_7 groups form a barrier to oxygen diffusion.

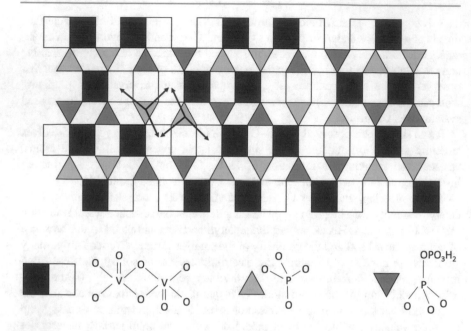

Fig. 4.2 Schematic of a complex surface of $(VO)_2P_2O_7$. The arrows represent the pathways for surface oxygen mobility, whereas the P_2O_7 groups constitute a barrier to oxygen diffusion. Adapted from Agashar et al. [4] and Vedrine [5]

4.2.1 Single-Site Heterogeneous Catalysts [1, 2]

These important catalysts, which were called single-site heterogeneous catalysts (SSHCs) by Sir John Meurig Thomas, are intrinsically superior to their homogeneous counterparts because i) they facilitate the separation of reaction products from reactants and ii) they can be readily reactivated by an appropriate chemical treatment.

Thomas et al. defined the "single site" as a catalytically active centre that is structurally well characterized and consists of one or more atoms; each single site is spatially isolated from the other sites and has the same energy of interaction with a reactant molecule [3]. Active sites are readily accessible to reactants due to their porous supports (zeolites, ordered silicas, etc.) with highly specific surface areas. Such "ideal" catalysts are highly desirable in heterogeneous catalysis since they give molecularly pure products and provide high selectivity for a wide variety of catalytic reactions (e.g., selective oxidations, selective hydrogenations, isomerizations, and oligomerizations). Therefore, SSHCs are inorganic analogues of enzymes. In fact, SSHCs must be prepared such that only one distinct type of active site is present, similar to any enzyme.

Since a fundamental assumption applicable to SSHCs is that the site is unique, the highest selectivity must be achieved, and the heat of the adsorption of the reactants must be the same regardless of the recovery rate. This finding suggests that as described in the literature, some SSHCs may not completely satisfy the definition proposed by Sir John Meurig Thomas. In fact, the authors often refer only to well-dispersed active sites on porous supports [5]. Nevertheless, the incorporation of transition metals into micro- and mesoporous supports represents an attractive strategy for preparing effective catalysts.

In summary, there are different types of SSHCs as follows:

(i)   Isolated active sites (e.g., ions, atoms, molecular complexes, and clusters) anchored to high-surface-area solid supports;
(ii) Immobilized asymmetric organometallic species on high-surface-area mesoporous supports;
(iii) "Ship-in-bottle" structures in which active sites are entrapped within zeolitic cages that are accessible to reactants;
(iv) Microporous materials in which the active sites are located at (or adjacent to) ions that replaced the framework ions of the parent structure.

A characteristic example is the titanium silicalite catalyst (TS-1), which is effective in several oxidation reactions with hydrogen peroxide, including the hydroxylation of benzene, the ammoximation of cyclohexanone, the epoxidation of olefins, and the oxidation of hydrocarbons and alcohols. TS-1 combines the advantages of isolated Ti^{4+} species with the hydrophobicity of the framework while retaining the spatial selectivity and specific local geometry of the isolated fourfold coordinated TiO_4 units of the molecular sieve structure.

The presence of framework tetrahedral Ti^{4+} sites can be confirmed by the IR spectra of TS-1 dispersed in KBr pellets. Indeed, a "fingerprint" band appears at 960 cm^{-1} due to the formation of Si–O–Ti bridges (Fig. 4.3A). Similarly, the $Ti^{4+}O^{2-} \rightarrow Ti^{3+}O^{-}$ charge transfer of isolated Ti^{4+} incorporated within the framework appears at approximately 210 nm in the (DR) UV–Vis spectrum (Fig. 4.3B).

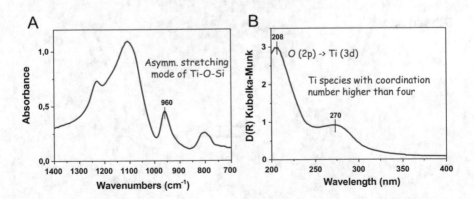

Fig. 4.3 FT-IR spectrum of the TS-1 catalyst dispersed in KBr pellets (section A) and (DR) UV–Vis spectrum of the same sample outgassed at 150 °C (Section B). Adapted from Piumetti et al. [1]

A further signal at approximately 270 nm indicates the presence of highly reactive Ti species with a CN higher than four but still isolated in the framework [1].

Self-repair in single-site heterogeneous catalysts [2]

An important step for reactions at solid surfaces is the self-repair of weakened (or disrupted) bonds at active sites once the catalytic cycle has concluded. To regenerate active sites, the displaced atoms must return to their original positions. Self-repair is an essential property of any effective catalyst and occurs locally at each active site. Self-repair is a form of self-organization occurring at catalytic surfaces.

An example is the epoxidation of linear olefins using the TS-1 catalyst and hydrogen peroxide. The following two different structures of isolated Ti^{4+} sites can be observed: the so-called perfect "closed" $Ti(OSi)_4$ tetrahedral site and the defective "open" structure $Ti(OSi)_3(OH)$. In contact with a single Ti^{4+} site, the reaction occurs via a peracid-like mechanism with an electrophilic cyclic structure as the active centre. A schematic of this reaction mechanism is shown in Fig. 4.4.

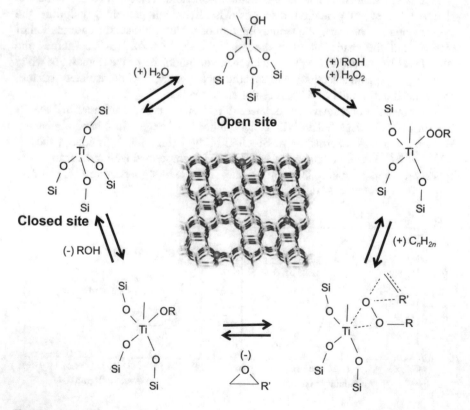

Fig. 4.4 Self-repair mechanism in an epoxidation reaction over the TS-1 catalyst. Adapted from Piumetti and Lygeros [2]

The marked ability of the Ti sites inside the silicalite framework to self-repair their local environment in a reversible (or nearly reversible) way upon interaction with adsorbates is probably the key to the reactivity of this Lewis catalyst [2].

4.2.2   Small Metal Particles [1, 2]

At the beginning of the 1980s, it was noted that gold, which is a well-known inert metal, may exhibit excellent catalytic activity at the nanoscale. The pioneering work by Haruta et al. [6] led to the discovery of the unique reactivity of Au nanoparticles dispersed over oxides (e.g., α-Fe_2O_3, Co_3O_4 and NiO) in numerous reactions, including the oxidation of CO and hydrocarbons, the oxidation-decomposition of amines and organic halogenated compounds, the hydrogenation of carbon oxides, unsaturated carbonyl compounds, alkynes and alkadienes, and the reduction of nitrogen oxides. Consequently, research efforts concerning Au-based nanocatalysts and metal nanoparticles in general have increased exponentially over the last few decades.

As a whole, a nanocluster consists of a small aggregate of up to 20–30 atoms, whereas a nanoparticle, which is in the size range of 3–10 nm, contains 10^3–10^7 atoms. Metal nanoparticles typically exhibit a continuous band of electronic energy levels and a well-defined Fermi surface, whereas small clusters have discrete energy levels and much higher HOMO–LUMO energy levels. Reactant molecules may exchange electrons to and from metal cluster surfaces with a lower energy expenditure due to their molecular character. These energetic considerations explain some fundamental differences in the catalytic properties of metal clusters and metal nanoparticles.

However, the properties of supported metal nanoparticles depend on their geo-metrical structure and surface features. Indeed, "real" supported metal catalysts typically consist of nanoparticles that expose different crystal planes, edges, corners and structural defects that interact differently with the reaction intermediates to yield different reactivities.

Typically, face-centred cubic (FCC) metals (e.g., Pt, Rh, and Au) tend to adopt a cuboctahedral structure, as this structure is usually associated with a minimum surface energy. However, many studies have examined the equilibrium shapes of Au nanoparticles in the range of 50–5000 atoms and found several shapes. Overall, small metal clusters are more reactive than larger nanoparticles because of the higher number of coordinatively unsaturated sites. Indeed, a decrease in the particle size increases the relative number of exposed corners (CN = 6) and edge sites (CN = 7), which, in turn, tends to increase the reactivity normalized to the number of exposed atoms.

The fraction of surface atoms only slightly changes when the particle size decreases from 10 to 4 nm (Fig. 4.5a). However, the percentage of exposed corner (highly reactive) and edge atoms significantly increases because the particle size is less than this value (Fig. 4.5b).

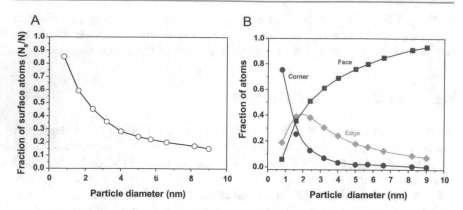

Fig. 4.5 Estimated fractions of surface atoms (N_S/N) with different metal nanoparticle sizes in which N_S and N are the number of surface atoms and the total number of atoms, respectively (Section A); trend of corner, edge and face atoms (normalized to exposed atoms) as a function of the particle diameter (Section B). Adapted from Piumetti et al. [1]

It is well established that low-coordination catalyst atoms are typically more reactive than high-coordination catalyst atoms. According to the concept of bond order conservation, for a single atom, the total binding capacity with the surrounding atoms is constant The proposed expression of the bond strength Q_n of a bond that has to be shared with n metal atom neighbours is as follows:

$$Q_n = \frac{1}{n}Q_0\left(2 - \frac{1}{n}\right)$$

where Q_0 is the bond strength when bonds do not have to be shared (i.e., n = 1). Hence, surfaces appear more reactive when surface metal atoms have fewer neighbours [7]. A schematic of the increase in the adsorption strength of a surface metal atom as a function of the metal atom CN is shown in Fig. 4.6.

This model is consistent with the notion that atoms located at the corners have a major effect on the activity of metal nanoparticles. However, decreasing the particle size to very small atomic ensembles can sometimes lead to lower activity [1].

According to Langmuir, *"the surface atoms are not rigid but tend to arrange themselves so that the total energy will be at a minimum. In general, this will involve a shifting of the positions of the atoms with respect to each other"*. Consequently, the inward relaxation and reconstruction of surfaces are much more marked at low-coordination sites (i.e., steps and corners), which are more "flexible" and can self-repair the catalytic surface at higher rates. Indeed, the faster the restructuring rate, the higher the reaction turnover rate, although it has also been observed that a rapid restructuring of surfaces may cause instability.

The equilibrium 3-D shape of metal nanoparticles can be determined by the Wulff rule, which posits that the convex envelope of planes (perpendicular to the

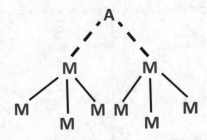 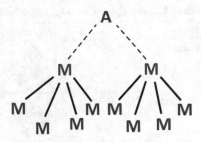

**Three metal atom neighbors
(More reactive surface)**   **Four metal atom neighbors
(Less reactive surface)**

Fig. 4.6 Schematic of adatom (A) chemisorption as a function of the metal coordination number. Adapted from van Santen [7]

surface normal) minimizes the surface energy of a given enclosed volume [2, 8]. The minimum surface energy of a given volume of a polyhedron is achieved when the distances of its faces from a given point are proportional to their specific surface energies as follows:

$$h_{hkl} = \lambda \gamma_{hkl}$$

where h_{hkl} represents the distance of the surface (hkl) from the centre of the nanoparticle, γ_{hkl} is the specific surface energy, and λ is a constant. Thus, for each considered (hkl) surface, a plane normal to the vector (hkl) is determined at the distance from the origin proportional to γ_{hkl}.

Depending on the surface energy of the facets and the adsorption bond strength of the molecules from the surrounding gas phase, nanosized metal particles typically form convex polyhedral structures with the Euler characteristic of sphere (X = 2). The Euler characteristic of polyhedra can be defined as follows:

$$X = V - E + F$$

where V, E, and F are the numbers of vertices, edges and faces in the given polyhedron, respectively. Therefore, convex polyhedra may have different fractions of corners, edges and terraces depending on their shapes as shown in Table 4.1 and Fig. 4.7. These structures are dominated by various crystallographic features with different surface atom densities, electronic structures, bonds, chemical reactivities and thermodynamic properties. Therefore, it is clear that changes in the particle shape and atomic arrangement due to different synthesis conditions may lead to different reactivities resulting from varied structures [2].

Table 4.1 Examples of convex polyhedra exhibiting different shapes and geometric properties

Name	3-D solid	Type	Classification
Cube (C)		Platonic solid	Regular Convex
Octahedron (O)		Platonic solid	Regular Convex
Icosahedron (I)		Platonic solid	Regular Convex
Cuboctahedron (CO)		Archimedean solid	Semiregular Convex
Truncated cube (TC)		Archimedean solid	Semiregular Convex
Truncated octahedron (TO)		Archimedean solid	Semiregular Convex
Rhombic dodecahedron		Catalan solid	Semiregular Convex
Decahedron (D)		Dipyramid	Semiregular Convex

Many studies have investigated the equilibrium shapes of FCC metal nanoparticles (i.e., Au, Pt and Pd) and showed several convex polyhedral structures, including icosahedron and truncated icosahedron, decahedron, truncated octahedron and cuboctahedron (or variants of these shapes), and singly or multiply twinned structures. However, it has been observed that real solid surfaces exhibit nonconvex regions; in fact, the presence of surface defects, irregularities and porosities leads to the formation of nonconvex polyhedra with various Euler characteristics ($X \neq 2$) as shown in Fig. 4.8. Furthermore, nonconvex structures, such as Kepler-Poinsot star polyhedral, may also form as a consequence of the selective passivation and blocking of edge and corner sites during growth coupled with anisotropic growth rates in different orientations. As a result, it is important to know the details of the surfaces of real (nano)catalysts, which are typically nonconvex structures, to study outcomes in terms of surface reactivity [2].

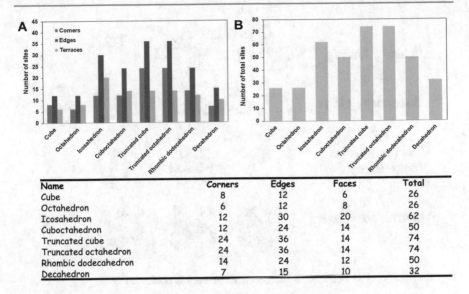

Name	Corners	Edges	Faces	Total
Cube	8	12	6	26
Octahedron	6	12	8	26
Icosahedron	12	30	20	62
Cuboctahedron	12	24	14	50
Truncated cube	24	36	14	74
Truncated octahedron	24	36	14	74
Rhombic dodecahedron	14	24	12	50
Decahedron	7	15	10	32

Fig. 4.7 Number of edges, corners and faces of convex polyhedra (Section A), number of total sites (Section B), and summary table. Adapted from Piumetti et al. [2]

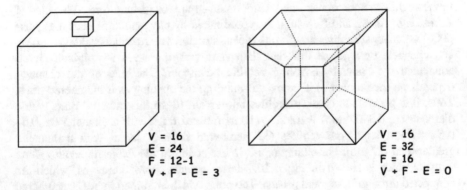

Fig. 4.8 Calculated Euler characteristics of a defective structure (left) and a porous solid (right)

4.2.3 Zeolites [2]

Zeolites are hydrated aluminosilicates with a three-dimensional framework structure constructed of SiO_4 and AlO_4 tetrahedra linked through oxygen atoms. Zeolites are flexible materials formed by self-assembled units with different structural levels (primary, secondary, tertiary and quaternary units), similar to protein structures (see Chap. 1) (Fig. 4.9).

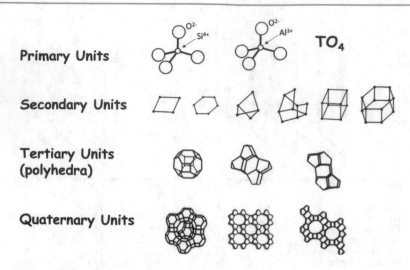

Fig. 4.9 Development of a zeolitic structure from primary to quaternary units

Thus far, more than 200 different zeolite frameworks have been identified, and over 40 naturally occurring zeolite frameworks are known. Most zeolites have a framework structure constructed from main-group element atoms (so-called T atoms, e.g., Al, Si, and P), which are coordinated by oxygen atoms, and may form [TO_4] tetrahedra. Zeolites are porous (open) structures that contain voids (cavities), i.e., channels or cages of different shapes. In fact, the cages are typically interconnected by channels, enabling zeolitic behaviour. The size of the channels depends on the number of T atoms surrounding the opening as n-membered rings. Thus, it is possible to classify zeolites into small- (6- to 8-membered rings with a diameter ca. 0.3–0.4 nm) medium—(10-membered rings with a diameter ca. 0.5–0.6 nm) and large-pore zeolites (12-membered rings or larger with a diameter greater than 0.7 nm). For example, a 3-D projection of the Faujasite zeolite along with its voids is shown in Fig. 4.10. These schemes show the cages, which are connected through hexagonal prisms. The pore, which is formed by a 12-membered ring, has a diameter of 0.7 nm.

The regular arrangement of the voids and their sizes, whose dimensions are of the same order of magnitude as molecular diameters, enable the zeolites to act as molecular sieves. Due to this outstanding property, zeolites are valuable as selective adsorbent solids for separating molecules and shape-selective catalysts.

Zeolites contain defects of various natures and active centres (e.g., acidic, basic, redox, or their combination) arranged with a certain nanoscopic (at the unit cell level) and macroscopic (at the crystal level) distribution. Accordingly, these materials can act as reaction channels, and their properties can be tailored by introducing transition metals and main-group elements (e.g., B, Ga, Fe, Cr, Ge, Ti, V, Mn, and Co) to the framework.

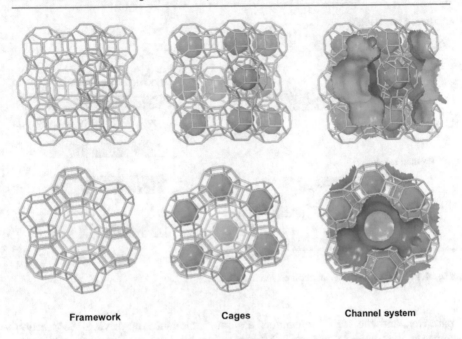

Framework   Cages   Channel system

Fig. 4.10 3-D projections of the Faujasite structure along with its cages (green) and channel system (light blue). Animated drawing software on the IZA-SC website was used to generate these images

Indeed, while pure silica frameworks are electrically neutral, the substitution of trivalent aluminium for tetravalent silicon imparts a negative charge to the material that must be balanced by positively charged extra-framework cations. Consequently, the presence of electric fields and controllable adsorption features within the pores allow the design of unique types of catalytic materials [2]. The reported surface electric fields of zeolites, such as mordenite, Beta, X, Y ZSM-5 and cation-exchanged zeolites, are in the range of 2–8 V nm^{-1} [9].

Furthermore, geometric changes in the ring may occur as a consequence of Al^{3+} substitution and H$^+$ addition. For example, Al substitution into Faujasite causes lattice deformations (relaxations) that are mainly accommodated by changes in the T-O-T bond angles (e.g., from 130° to 160°).

Due to their microporous nonconvex framework, zeolite-type materials are stereoselective and behave similarly to enzymes. In fact, these materials can discriminate among different molecules, giving rise to molecular diffusion within a complex channel system. Thus, the following three different types of shape selectivity may appear within a zeolite-type material as shown in Fig. 4.11: (i) reactant selectivity, (ii) product selectivity and (iii) transition-state selectivity. In particular, the last type occurs when the specific arrangement of the solid is such that only certain configurations are effective for the transition state. Therefore, the framework geometry in the neighbourhood of the active site can strongly influence the reaction

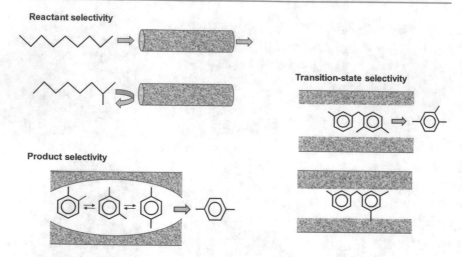

Fig. 4.11 Stereoselectivity in zeolite-type materials

pathway, and the steric constraints assume importance in driving the reaction towards the desired products. Typical examples of transition-state selectivity include the cracking of paraffins and the inhibition of coke formation within the pores of ZSM-5 crystals.

Zeolites are typically hydrophilic materials that can be relatively selective adsorbents for polar or nonpolar compounds depending on the framework Si/Al ratio and the nature of any extra-framework species. Therefore, the polarity of a given zeolite can be tuned by varying the Si/Al ratio.

Generally, when a guest molecule has a polar character, electrostatic interactions with the zeolitic framework cannot be neglected. Specifically, when a molecule is confined in the cavities of a zeolite, the sorption energy depends on different terms. Indeed, the electric field and its potential play a significant role in determining the adsorption and conversion of hydrocarbons within the zeolite framework as observed in the case of the cracking of paraffins on several aluminosilicates. The propensity of field gradients to weaken some hydrocarbon bonds depends on the charge density (i.e., aluminium content), zeolitic structure, temperature and polarizing effect of cations.

Zeolites can be viewed as acidic catalysts. In fact, zeolites usually contain Brønsted acid OH groups (namely, $SiO(H)Al(OSi)_3$), which can function as active sites capable of activating chemical bonds (Fig. 4.12).

The acidic strength of sites can be modified by the isomorphic substitution of Si for trivalent atoms other than Al. For example, Ga- and B-substitution can produce tailored zeolites with stronger and weaker Brønsted sites, respectively. Moreover, the framework types and crystallographic positions of the sites also play important roles in defining the acidity of zeolites.

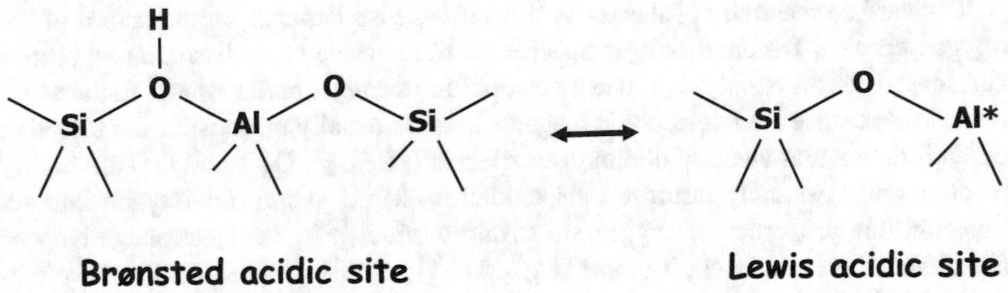

Brønsted acidic site   **Lewis acidic site**

Fig. 4.12 Schematic of Brønsted and Lewis acidic sites in a zeolite framework

Zeolite acidity is critical for specific catalytic reactions, e.g., in the catalytic cracking of alkanes and alkenes. In fact, the reaction network occurring at a high temperature is acid catalysed and proceeds through the formation of various carbocation transition states; because of the feedstock complexity and the intricate reaction pathways, the number of possible products and by-products is large, and selectivity towards specific compounds can be enhanced by tuning the zeolite acid sites [2].

4.2.4   Oxide Catalysts [2]

These catalytic systems exhibit self-organization since a dynamic network of joined sites may promote oxidation reactions. This redox-cycle mechanism occurs quite frequently in oxidation reactions over vanadium oxide catalysts as originally described by P. Mars and D. van Krevelen (Fig. 4.13).

However, evidence suggests that other metal oxides, such as those of Mo, Cr, and Fe, can also operate via redox cycles. The basic requirement for the MvK-type mechanism is a facile redox behaviour of the supported metal (e.g., V, Cr, and Mo) and good mobility of both electrons and lattice oxygen anions within the framework. The other factors include the facile activation of reactant molecules, the presence of surface defects, the surface acid–base properties and the temperature.

Fig. 4.13 C-H bond activation over a vanadium oxide catalyst. Adapted from Piumetti et al. [1]

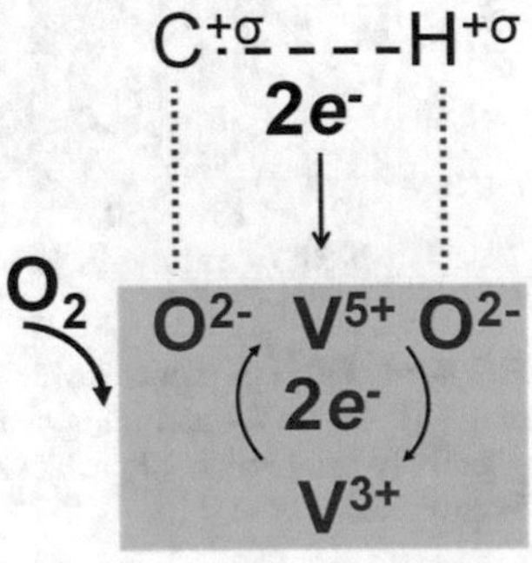

The surface reactivity of metal oxide catalysts also depends on the nature of the oxygen species. Different oxygen species on the surface have dissimilar stabilities and may react differently, e.g., via electrophilic or nucleophilic attack. Lattice O^{-2} ions at the surface are nucleophilic reagents and are usually responsible for selective oxidation, whereas electrophilic oxygen species (i.e., O_2^{-}, O_2^{-2} and O^{-}) are highly reactive and eventually result in total oxidation. Moreover, metal–oxygen interactions and the properties of oxygen are strongly affected by the electronegativity of the element present in the support (e.g., Al, Ti, or Si). Consequently, the effectiveness of active species may be largely influenced by the nature of the support. The complexity of these oxide catalysts acting in a dynamic network is further influenced by the fact that acid–base and redox properties may be closely related to each other. For instance, it has been observed that the oxidative dehydrogenation (ODH) of propane to propene over VO_x-SiO_2 catalysts also depends on the Brønsted acidity of V species as measured by IR spectroscopy. In fact, the best catalysts (in terms of selectivity to propene) are characterized by the presence of weaker Brønsted acidic sites (V–OH groups) as shown in Fig. 4.14. However, higher acidity promotes the formation of total oxidation products rather than propene [10].

Therefore, several factors affect the performance of oxide catalysts, and ensembles of atoms (ions) and electrons mutually interact in an active network [1].

Moreover, several studies revealed that surface redox cycles on transition metal oxides typically occur at moderate temperatures (i.e., from 300 to 500 °C) and are replaced at higher temperatures by radical processes in the gas phase. However, below 100 °C, oxidation reactions may occur through peroxidic mechanisms that

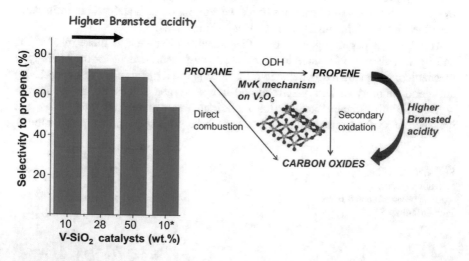

Fig. 4.14 ODH of propane over VO_x-SiO_2 catalysts prepared by flame pyrolysis (V content in the range 10–50 wt.%) and impregnation procedure (asterisk). The results are expressed in terms of selectivity to propene (%) achieved at propane isoconversion (ca. 12% propane ± 1%). Adapted from Piumetti et al. [10]

directly involve adsorbed radicals. A "transitional region" in the temperature range from 100 to 300 °C has been proposed in which complex surface oxidation mechanisms that should be responsible for the formation of total oxidation products occur. This pattern reflects the high complexity of such oxide catalysts [2].

4.3 Active Sites in Enzymatic Catalysis

Due to its amino acid distribution, the active site of an enzyme is uniquely designed in terms of its size, shape and chemical behaviour to bind specific targets (a single substrate or more) and promote a biochemical reaction.

As previously discussed in Chap. 2, the main properties of the binding-active sites in enzymes are as follows:

- Complementarity: Molecular recognition depends on the 3-D structure of the enzyme.
- Flexibility: The 3-D structure allows enzymes to adapt to ligands.
- Adaptive surfaces: Binding sites can be concave, convex or flat surfaces. Rough (or even fractal) surfaces are possible.
- Noncovalent forces: Binding sites are characterized by weak interactions.
- Affinity: Binding ability between an enzyme and a substrate. Competition can exist among different substrates for the same binding site of an enzyme.

4.3.1 Allosterically Regulated Enzymes: The Case of ATCase

Aspartate transcarbamoylase (ATCase) is an allosterically regulated enzyme which possesses a unique quaternary structure involving separable catalytic and regulatory subunits. ATCase consists of two catalytic trimers (with catalytic activity) and three regulatory dimers (with regulatory functions) that are completely separable units (Fig. 4.15). ATCase is largely α-helical, and the quaternary structure is highly flexible; thus, a significant change in the quaternary structure may occur during catalysis.

ATCase catalyses the initial step in the biosynthesis of pyrimidines, which finally yields pyrimidine nucleotides, like cytidine triphosphate (CTP). The rate of the reaction catalysed by ATCase depends on the CTP concentration: it is fast at low [CTP] and slows down as [CTP] increases. High CTP levels inhibit ATCase by binding an allosteric/regulatory site. As discussed in Chap. 2, allosterically regulated enzymes, such as ATCase, are different from traditional enzymes because they are not only affected by the substrate concentration but also regulated by other compounds [11, 12].

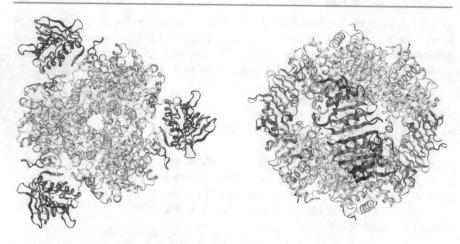

Fig. 4.15 Different views of aspartate transcarbamoylase from PDB code 8ATC. This scheme depicts catalytic trimers (yellow chains) and regulatory dimers (orange-red chains)

4.3.2 Active Sites and Electric Fields

Linus Pauling and early biochemists [13, 14] believed that enzymes strain substrates along reactive coordinates towards their transition states. This hypothesis was subsequently labelled ground-state destabilization (GSD) and was supported by Phillips et al. [15] who studied the structure of lysozyme. Pauling also argued that enzymes stabilize their reaction transition states, and the theory of the transition-state stabilization (TSS) of enzyme catalysis is currently accepted. However, some results obtained using catalytic antibodies suggest that TSS does not provide a complete model for understanding enzyme catalysis [16].

Enzymes do not act simply by placing reacting groups of substrates in close proximity. In fact, statistical mechanics suggests that such an effect could raise $\Delta S^{\ddagger}$ by 15–18 cal K^{-1} mol^{-1} (or a rate improvement of 10^4–10^5), and this variation is too small to account for the much higher rate enhancements that enzymes usually achieve [16].

Page and Jencks [17] argued that enzymes reduce the substrate entropy linked to rotations and internal degrees of freedom. This idea was experimentally confirmed in experiments performed by Bruice and Pandit [18].

Chemical reactions usually need rearranging atoms in molecules to occur. In the case of polar reactions, a reactant's dipole is expected to change on activation, which implies that the transition state can be stabilized by solvent reorganization. Thus, when a molecule (solute) with a permanent dipole is placed in a solvent (e.g., water), the dipoles of the solvent molecules realign in the same direction to the solute's dipole, so that the overall potential energy of the system can be minimized (Fig. 4.16). Then, a net solvent reaction field $\overrightarrow{F}_{solv}$ occurs on the dipole aligned in the same direction as $\overrightarrow{\mu}_{solute}$ (the solute's dipole moment) at equilibrium conditions.

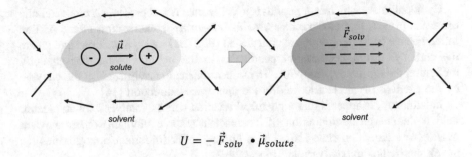

$$U = - \vec{F}_{solv} \bullet \vec{\mu}_{solute}$$

Fig. 4.16 Schematic of a solute molecule with a permanent dipole $\vec{\mu}_{solute}$ that organizes polar solvent molecules (e.g., water molecules) around it so that they exert a net solvent reaction field $\vec{F}_{solv}$ on the permanent dipole aligned in the same direction as $\vec{\mu}_{solute}$. Adapted from Fried and Boxer [16]

This interaction leads to the interaction energy U, which can be calculated with the following formula:

$$U = - \vec{F}_{solv} \cdot \vec{\mu}_{solute}$$

The reaction field points in the same direction as the dipole of the solute. Its magnitude depends on the solute's dipole and the solvent's polarity [16].

The electric field results from the dipole–dipole interactions, the dipole-induced dipole interactions, and the hydrogen-bonding interactions provided by the solvent molecules. For typical solvents, the electric field ranges from 0 to 10^9 V m^{-1}.

Thus, electric field catalysis may explain the catalytic strategy used by enzymes acting in solvent that stabilizes the dipole moment of the transition state via electric fields.

Briefly, electric field catalysis describes the catalytic reactions within an environment that stabilize the transition state's dipole more than the reactant's dipole as follows:

$$\Delta \Delta G^{\ddagger} = - \left[\left(\vec{F}_{env,TS} \cdot \vec{\mu}_{TS} \right) - \left(\vec{F}_{env,R} \cdot \vec{\mu}_{R} \right) \right]$$

where $\vec{\mu}_R$ is the dipole moment of the reactant, $\vec{\mu}_{TS}$ is the dipole moment of the transition state, $\vec{F}_{env,R}$ is the electric field exerted by the environment on the reactant dipole, and $\vec{F}_{env,TS}$ is the electric field exerted on the transition state dipole by the environment.

Although this approach may appear complex for a whole molecule, it is possible to focus on substrates in which charge rearrangement occurs primarily at a single site (e.g., a carbonyl group) and express the local electric field at a carbonyl bond as $\vec{F}_{env}^{C=O}$.

Cyclophilin A, which is a peptidyl-prolyl isomerase, represents a characteristic example. In fact, this enzyme catalyses the *cis–trans* isomerization of the proline imide peptide bond in only one chemical step (Fig. 4.17) via a transition state where the carbonyl group becomes perpendicular to the peptide plane, consequently realigning the dipole by about 90°. The transition state is stabilized by the enzyme by the exertion of an electric field in the appropriate direction [16].

In summary, it appears that a chemical reaction can be catalyzed by an electric field if the charge configuration of the reactant (dipole moment) changes when passing to a transition state [16]. These findings offer important opportunities for novel approaches to catalysis and biocatalysis.

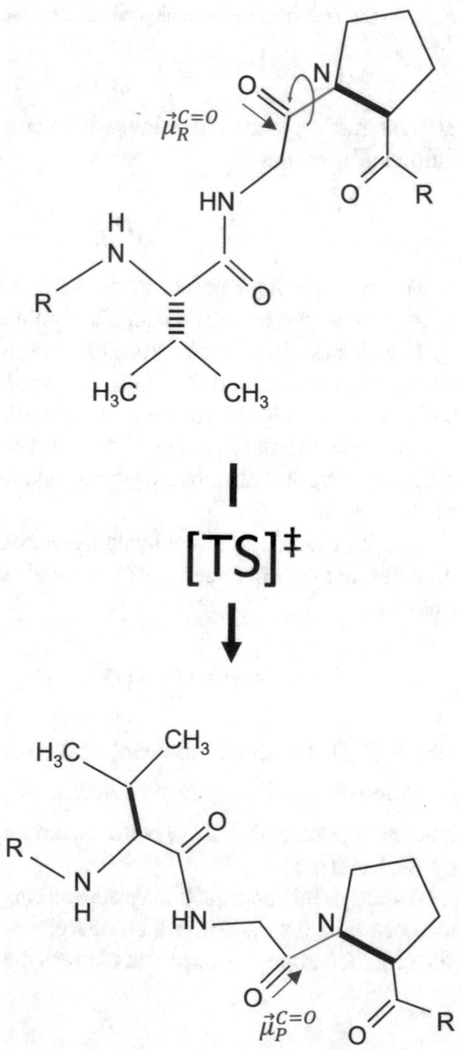

Fig. 4.17 Effect of orientational electric field catalysis: cyclophilin A rotates the N-terminal section of the Gly-Pro peptide bond. Adapted from Fried and Boxer [16]

4.4 Active Sites in Homogenous Catalysis

Homogeneous catalysis is single-site and molecular in character. The spatial separation of the active centres from each other and the high degree of control of the
site structure as well as its environment are the principal advantages of homogeneous catalysis. Several types of homogeneous catalysts are known, including the
following:

- Soluble acids and bases (frequently applied in organic chemistry);
- Soluble organic catalysts (organocatalysis);
- Homogeneous enzyme catalysts;
- Homogeneous transition metal catalysis.

In the last class of homogeneous catalysts, the metal centre represents the active
site and is surrounded by ligands.

Transition metals are particularly interesting as homogenous catalysts since they
have free d and f orbitals that are able to overlap with the orbitals of unsaturated
organic compounds and ligands. Moreover, these metals can exhibit different
oxidation states and can usually easily change their oxidation state. Transition metal
complexes appear highly versatile and contain one metal (M) at the least, and one or
n ligands (L), i.e., ML_n.

Therefore, the empty d and/or f orbitals of the transition metal can be filled by
electron-donating ligands, and form a σ-bond to the metal centre. In this situation,
the metal operates as an electron acceptor. However, the transition metal also
possesses filled orbitals able to donate electrons to form π bonds with ligands, such
as hydrogen, halogens, organic phosphorus, oxygen, sulfur, and nitrogen compounds. For instance, in Zeise's salt $K[PtCl_3(\eta^2\text{-}C_2H_4)]$, ethylene is bound side-on
to Pt, where there is an overlap between the filled π orbital of ethylene with an
empty orbital of platinum. Moreover, a filled d orbital of Pt overlaps with the
empty π^* orbital of the ethylene molecule with π back donation as shown in
Fig. 4.18 [18].

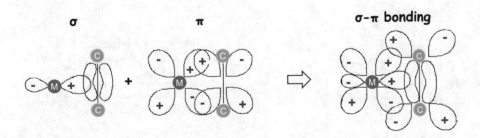

Fig. 4.18 Orbital status in the bonding of a metal with ethylene. Adapted from Beher [18]

4.4.1 Catalytic Cycles and Flexibility

A homogeneously catalysed reaction occurs via elementary steps that form a catalytic cycle as shown schematically in Fig. 4.19. The catalytic active metal (M) first adds a reactant (R) and then a further reactant (R'). Both are then activated by the catalyst and interact with one another. The product (R-R') is obtained, and the metal is released again to start a new catalytic cycle. Nevertheless, the catalyst must be very flexible, i.e., must easily change its oxidation state in order to progress through all these steps. Figure 4.19 also shows the Rh-catalysed hydrogenation of an alkene. The reaction starts with active Rh(I) in the form of $[RhClL_2]$ (L = PPh_3, namely, triphenylphosphine), which initially adds a hydrogen molecule, thus giving Rh(III) dihydrido species. Then, the complex $[RhH_2ClL_2]$ is able to add the alkene, and at the metal, one hydrogen atom reacts with the alkene to form an alkyl species. However, there is no change in the oxidation state (+3) of Rh. In the final step, the alkyl substituent reacts with the second hydrogen atom to form an alkane. Thus, the cycle ends, the starting Rh(I) complex is recovered, and a new cycle can be performed [18].

The flexibility of transition metals can also be observed in the variation in their CNs and geometry during the catalytic cycle. A characteristic example is the hydrogenation of ethylene by an Rh(I) complex (Fig. 4.20).

The catalytic cycle starts with a Rh(I) complex with CN 4. Then, this complex interacts with a hydrogen molecule, forming an octahedral complex (CN 6). As there are no more available coordination sites on this complex, one ligand L must be split off to give a pyramidal species with CN 5. Then, ethylene can coordinate to Rh, forming octahedral coordination (CN 6). A pyramidal pentacoordinate alkyl complex is obtained (CN 5) by the insertion of ethylene into one of the Rh–hydrido bonds. Hence, the product ethane is released, and a trigonal-planar complex is formed (CN 3). This new complex can either directly add H_2 to form pyramidal species (CN 5) or firstly add the ligand and then H_2 [18].

The catalytic active site in homogeneous catalysts is not usually well understood. In certain cases, catalytically active sites are created in situ starting from a metal salt and some ligands. Thus, the term "catalyst precursors" appears appropriate given that the precise nature of the active species in homogeneous catalysis often remains unknown.

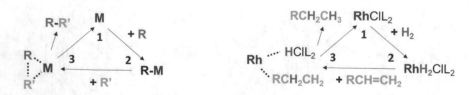

Fig. 4.19 Scheme of a catalytic cycle (left) and oxidation states of the catalytically active metal during the Rh-catalysed hydrogenation of an alkene (right). Adapted from Beher [18]

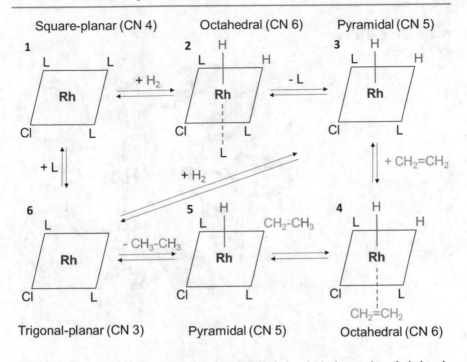

Fig. 4.20 Coordination number and geometry of Rh(I) during the hydrogenation of ethylene by an Rh-containing complex. Adapted from Beher [18]

For instance, the Wilkinson catalyst [RhCl(PPh₃)₃] is not the active hydrogenation catalyst as its three sizeable triphenylphosphine ligands do not allow the coordination of large alkenes (Fig. 4.21).

Indeed, the first step in a Wilkinson-type homogeneous reaction is the ligand exchange of a triphenylphosphine with a small solvent molecule (e.g., tetrahydrofuran or toluene) (Fig. 4.22). This solvent molecule is bonded weakly and can be easily replaced by other compounds. Then, H_2 coordinates to Rh in an oxidative addition step. Rh(III) is obtained, and an octahedral structure (CN 6) is formed. The weakly bound solvent molecule dissociates with the appearance of a more strongly coordinating alkene. The alkene inserts into a Rh–H bond by migratory insertion and forms Rh alkyl species. The sixth coordination structure is accessible and can be occupied by a solvent molecule. The final stage is the reductive elimination of the alkane by the linking of the alkyl substituent with the second hydride. Finally, the active Rh(I) is re-established, and a new catalytic cycle may start [18].

Examples of redox-type mechanisms can be observed in homogeneous oxidation catalysis, e.g., in the Wacker-type process, which typically consists of the conversion of ethylene into acetaldehyde by oxygen in water in the presence of Pd and Cu species. In fact, in the Wacker system, Pd^{2+} is the active intermediate that generates atomic oxygen from water as follows:

Fig. 4.21   Structure of the
Wilkinson catalyst [RhCl
(PPh$_3$)$_3$]

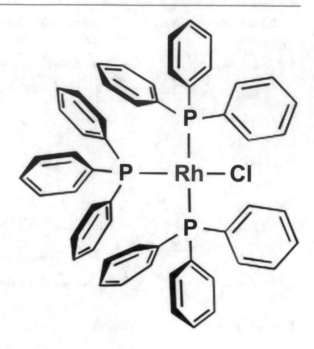

$$Pd^{2+} + H_2O + H_2C = CH_2 \rightarrow H_3C - CHO + Pd^0 + 2H^+$$

The following Cu-redox cycle is necessary for regenerating the Pd^{2+} species:

$$O_2 + 2Cu^+ \rightarrow 2Cu^{2+}O$$
$$2Cu^{2+}O + Pd^0 \rightarrow 2Cu^+ + Pd^{2+}$$

Therefore, the mechanism of this homogeneous Wacker reaction catalysed by Pd^0/Pd^{2+} and Cu$^+$/Cu^{2+} couples is similar to that of an MvK-type reaction (heterogeneous catalysis) or a ping-pong mechanism (enzyme catalysis) [19]. These findings suggest the unification of heterogeneous, homogeneous and enzymatic catalysis into one conceptually coherent model.

4.5   Summary

- Both geometric and electronic properties are fundamental factors for catalysis. The latter are often considered separately, although their interrelation in many catalytic reactions has been well established.
- Sir J. M. Thomas introduced the concept of SSHC and defined the "single site" as a structurally well-characterized catalytically active centre consisting of one or more atoms. Each site is spatially isolated from the other sites and has the same

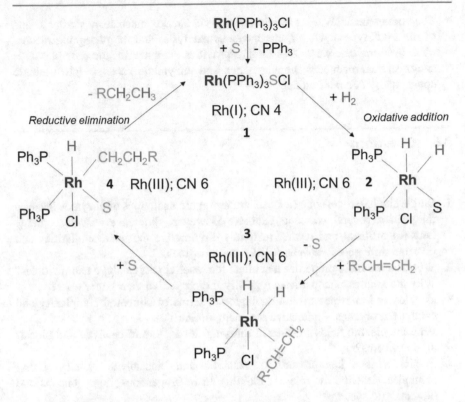

Fig. 4.22 Catalytic cycle of alkene hydrogenation with a Wilkinson-type catalyst (Solvent = S). Adapted from Beher [18]

energy of interaction with a reactant molecule. SSHCs are the inorganic analogues of enzymes.

- Metal-supported catalysts typically consist of nanoparticles and clusters exhibiting different shapes and geometric properties.
- Zeolite-type materials are stereoselective and behave similarly to enzymes with an interplay between electric fields and nanoporous structures. Thus, the electric field and its gradient play a role in determining the adsorption and catalytic conversion of hydrocarbons within the zeolite framework.
- Many oxidation reactions promoted by metal oxide catalysts occur via Mars-van Krevelen-type mechanisms in which the active sites are not isolated species but ensembles of mutually interacting atoms (ions) and electrons in a self-organized dynamically active network.
- Electric fields can be considered unifying descriptors for different catalytic systems.

- The homogeneous Wacker reaction occurs via a redox mechanism similar to that of the MvK-type reaction (heterogeneous catalysis) and the ping-pong mechanism (enzyme catalysis). This finding provides evidence for the possible unification of heterogeneous, homogeneous and enzymatic catalysis into a single conceptually coherent model.

4.6   Questions

1. Single site heterogeneous catalysts are inorganic analogues of enzymes because they share several common features. However, differences usually emerge between artificial and natural materials. Summarize the main similarities and possible differences between SSHCs and enzymes.
2. What are the main properties that affect the reactivity of catalytic nanoparticles? Why are smaller nanoparticles typically more reactive than larger ones?
3. Zeolites and enzymes exhibit analogies in terms of structural complexity and catalytic properties. Summarize the main similarities.
4. Why are electric fields considered unifying descriptors of catalytic and biocatalytic systems?
5. Flexibility is a key property of catalytic and biocatalytic systems. Using examples, describe the role of flexibility in heterogeneous, homogeneous and bio-catalytic systems.

References

1. M. Piumetti, F.S. Freyria, B. Bonelli, Chimica Oggi—Chem. Today **31**, 55–58 (2013)
2. M. Piumetti, N. Lygeros, Chimica Oggi—Chem. Today **31**, 48–52 (2013)
3. J.M. Thomas, R. Raja, D.W. Lewis, Angew. Chem. Int. Ed. **44**, 6456–6482 (2005)
4. P.A. Agashar, L. De Caul, R.K. Grasselli, Catal. Lett. **23**, 339–351 (1994)
5. J.C. Vedrine, Appl. Catal. A: General **474**, 40–50 (2014)
6. M. Haruta, Catal. Today **36**, 153–166 (1997)
7. R.A. van Santen, *Modern Heterogeneus Catalysis* (Wiley-VCH, 2017), pp. 83–84
8. M. Piumetti, M. Hussain, D. Fino, N. Russo, Appl. Catal. B: Environ. **165**, 158–168 (2015)
9. C.J. Rhodes, Chem. Papers **70**(1), 4–21 (2016)
10. M. Piumetti, M. Armandi, E. Garrone, B. Bonelli, Micropor. Mesopor. Mat. **164**, 111–119 (2012)
11. J.M. Berg, J.L. Tymoczko, L. Stryer, *Biochemistry. Intl*, 7th edn. (W.H. Freeman and Company, New York, 2012), pp. 300–306
12. H.M. Ke, W.N. Lipscomb, Y.J. Cho, R.B. Honzatko, J. Mol. Biol. **204**, 725–747 (1988)
13. L. Pauling, Chem. Eng. News **24**(10), 1375–1377 (1946)
14. J.B.S. Haldane, *Enzymes* (Longmans Green, London, 1930)
15. C.C.F. Blake, D.F. Koenig, G.A. Mair, A.C.T. North, D.C. Phillips, V.R. Sarma, Nature **206**, 757–761 (1965)
16. S.D. Fried, S.G. Boxer, Annual. Rev Biochem. **86**, 387–415 (2017)

17. T.C. Bruice, S.J. Benkovic, Biochemistry **39**, 6267–6274 (2000)
18. A. Beher, Catalysis, homogeneous, in *Ullmann's Encyclopedia of Industrial Chemistry* (Wiley-VCH Verlag GmbH & Co. KGaA, Weinheim, 2012), pp. 223–269
19. R.A. van Santen, M. Neurock, *Molecular Heterogeneous Catalysis* (Wiley-VCH, 2006), p. 474

Complexity in Catalysis

5

Marco Piumetti and Nik Lygeros

5.1 Self-organizing Systems [1]

Self-organization in chemical systems reflects the formation of ordered structures due to local interactions among the components in both thermodynamically closed and open systems. The former tend to reach equilibrium conditions by lowering their free energy, although kinetic constraints may limit this process. In contrast, open systems far from equilibrium exhibit an interplay between chemical reactions and transport phenomena. These systems undergo irreversible processes that create entropy and can be described by kinetic equations with broken time symmetry instead of canonical equations [1, 2].

Such nonequilibrium systems were first studied by I. Prigogine, who named these systems dissipative structures [3, 4]. The evolution of these systems can be described by nonlinear differential equations even if it is not possible to predict the exact phenomena occurring in systems far from equilibrium due to the fundamental limitations of the mathematical structures [5, 6].

Experimental studies investigating oscillatory kinetics have been performed with electrochemical reactions and subsequently the Belousov-Zhabotinsky (BZ) reaction in homogeneous systems, resulting in the establishment of nonlinear chemical oscillators. Subsequently, in the early 1970s, it was found that heterogeneous catalytic reactions may also exhibit oscillatory kinetics. For instance, it was revealed that CO oxidation over Pt-based catalysts may exhibit temporal oscillations (in terms of oxidation rates) depending on the operating conditions. Such effects of CO oxidation over Pt single-crystal surfaces were further investigated by G. Ertl. It was observed that under specific operating conditions (i.e., p_{CO} and T), a Pt (110) plane can alternate between two conditions (i.e., 1×2 and 1×1 structures), thereby giving rise to regular oscillations in the reaction rate (r_{CO_2}) [2]. This

Electronic Supplementary Material The online version of this chapter (https://doi.org/10.1007/978-3-030-88500-7_5) contains supplementary material, which is available to authorized users.

is a dissipative system with dynamic regimes that can be kinetically described by coupled first-order nonlinear differential equations (i.e., Eq. Lotka-Volterra equations) and their solutions for chosen parameters (Fig. 5.1) as follows:

$$\frac{dx}{dt} = a_1 x - a_2 xy$$
$$\frac{dy}{dt} = b_1 xy - b_2 y$$

The Lotka-Volterra equations describe a predator–prey model with oscillations in the population sizes of both the predators and prey, and the peak of the predator's oscillation lags slightly behind the peak of the prey's oscillation. Thus, the oscillatory trends of species x and y are coupled to each other with a phase shift.

In the case of CO oxidation on Pt (110), x, y corresponds to the surface coverages of O and CO. Since the concentration variables depend on space and time coordinates, spatiotemporal pattern formation typically occurs on the catalyst surface. Similarly, kinetic investigations of N_2O decomposition over Cu-containing ZSM-5 catalysts have shown reaction rates exhibiting isothermal oscillatory behaviours depending on the operating conditions and the nature of the active sites (vide infra) [7].

Oscillatory behaviours in biochemical systems have been observed in competitive inhibition, such as glycolysis or the enzymatic control of certain metabolic processes. The metabolism of a cell involves a complex network of reactions structured according to various metabolic pathways (e.g., convergent, divergent,

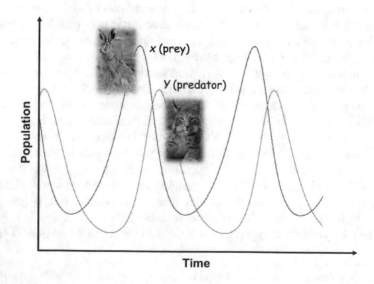

Fig. 5.1 The Lotka-Volterra (or predator–prey) model describing catalytic systems exhibiting temporal oscillations

cyclical and even polycyclic) and is activated by the presence of enzymes. Therefore, the metabolic processes occurring in living organisms are self-organized, and the metabolic pathways form ordered structures [8].

In multistep processes (e.g., glycolysis), some reactions occur under steady-state conditions; the reaction rates rise or fall depending on the substrate concentration. Other reactions are far from equilibrium conditions; the latter are typically the regulation points of the overall pathway. Nevertheless, the coexistence of multiple steady states leads to nonlinear effects, i.e., biochemical oscillations at the sub-cellular level.

5.2   Complexity of Catalytic Processes [9, 10]

Catalytic processes occurring under steady-state flow conditions can be described by a set of nonlinear partial differential equations combining the chemical kinetics with the diffusive phenomena of the adsorbed species as follows:

$$\frac{\partial x_i}{\partial t} = F_i(x_j, p_k) + D_i \nabla^2 x_i$$

where x_i corresponds to the state variables of the reactants (e.g., concentrations); F_i are the nonlinear operators expressing the chemical kinetics; D_i represents the diffusion coefficients; and p_k indicates a set of parameters. The solutions of such reaction–diffusion equations lead to the formation of self-organized patterns, which is a way to create fractal objects.

These systems can be unpredictable due to their intrinsic complexity, and original structures with radically different behaviour from the classical structure may appear. Indeed, there are unexpected relationships between the chemical kinetics and transport phenomena of open systems far from equilibrium. Consequently, such systems cannot be described by classical dynamics or quantum mechanics using deterministic, time-reversible approaches. This issue also explains their intrinsic complexity.

Self-organizing phenomena in catalysis, including transport phenomena, collective and dynamic behaviours, self-repair mechanisms of active sites and cooperative-synergistic effects, may reflect the presence of complex structures (Fig. 5.2) [9, 10].

These complex structures, which appear in both catalysis and biocatalysis, may lead to different microscopic, mesoscopic and macroscopic phenomena as shown in Fig. 5.3.

Since catalytic behaviour typically depends on nonlinear combinations of several complex structures, the performance of multicomponent catalysts differs from the sum of factors. Both the physicochemical properties of the catalyst and the chemistry of the process (e.g., reaction steps, pathways, etc.) affect the complexity of the catalytic reaction. For example, a reaction with a multistep character,

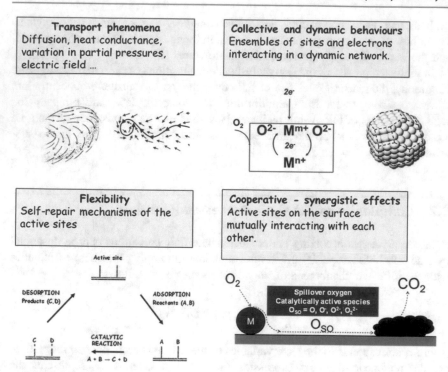

Fig. 5.2 Some examples of complex structures occurring in heterogeneous catalysis. Adapted from Piumetti and Lygeros [9]

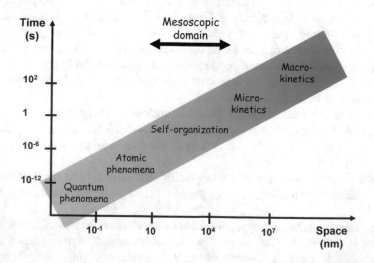

Fig. 5.3 Microscopic, mesoscopic and macroscopic dynamics occurring in catalytic processes. Adapted from the lecture "*Molecules at surfaces and mechanism of catalysis*" given by *G. Ertl*

including the generation of different intermediates, leads to high chemical complexity. However, the geometric and electronic properties of a catalyst may affect the structural complexity of the catalytic system. As a result, the overall complexity of a catalytic reaction (λ) can be described through a nonlinear combination (hyperoperation, vide infra) of chemical complexity (C_c) and structural complexity (C_s) as follows:

$$\lambda = C_c \circ C_s$$

5.2.1 Thermodynamic Considerations [9]

There is a fundamental difference between the dynamic and thermodynamic states of a catalytic system. To describe the dynamic state, it is necessary to have precise knowledge of the position (and motion) of the molecules that compose the whole system (knowledge of the microstate). Conversely, the thermodynamic state is determined from macroscopic properties (e.g., temperature, pressure, volume and density). According to statistical mechanics, a macrostate is represented by a probability distribution of states across a certain statistical ensemble of all microstates ($\pi_1, \pi_2, ... \pi_n$). Thus, a relationship exists between the number of microstates (π) and entropy S

$$S = f(\pi)$$

as formulated by Boltzmann:

$$S = k_B \log \pi$$

where k_B is the Boltzmann constant corresponding to the ratio between the gas constant and Avogadro's number (it has the same dimension as entropy). Therefore, for a given set of macroscopic variables, entropy measures the degree to which the probability of the system is spread across possible microstates. Thus, entropy reflects the number of ways in which a system may be arranged.

Therefore, if we consider two dynamical states as follows:

$$S_1 = f(\pi_1); S_2 = f(\pi_2)$$

The entropy of the system is

$$S < S_1 + S_2$$

and the probability of the system is the product of the two probabilities:

$$\pi = \pi_1 \pi_2$$

Then,

$$f(\pi_1 \pi_2) < f(\pi_1) + f(\pi_2)$$

Assuming that the complexity λ of the system is a function of dynamical states, a relationship exists between entropy S and complexity λ

$$\lambda$$

$$= g(S)$$

Therefore, complexity is a state function.

Dissipative systems exhibit spontaneous self-organization by releasing energy into the environment to compensate for their decreases in entropy. The occurrence of self-organizing phenomena tends to spontaneously organize the systems structurally, causing them to become progressively more ordered and differentiated. Thus, the energy for pattern formation is minimized, and self-organization (SO) occurs as the reverse of entropy production.

According to Shannon's formula [11], SO is given by

$$SO = 1 - \frac{S}{S_{max}}$$

where S/S_{max} corresponds to the ratio between entropy and its maximum value (Fig. 5.4).

Fig. 5.4 Relationship between self-organization and entropy in a dissipative system. Adapted from Piumetti and Lygeros [9]

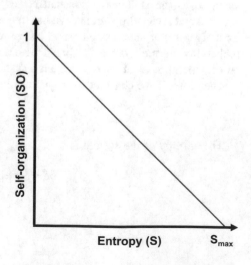

When the system has maximum entropy ($S \approx S_{max}$), *SO* is close to zero (chaotic regime). However, if the elements are ordered in such a way that given one element, the position of other elements is determined, the system's entropy S vanishes to zero. Thus, in a self-organizing system, the "internal order" increases over time, and *SO* becomes unity, indicating a complex system with pattern formation. As a result, an inverse correlation exists between entropy and complexity [9].

For example, biological systems tend to spontaneously organize themselves into complex differentiated structures, indicating that locally, they minimize entropy. Indeed, biological components (e.g., proteins, DNA, nuclei, cells, tissues, organs, etc.) are characterized by the spontaneous formation of anisotropic, ordered and complex structures that are rich in information but poor in entropy. As a result, catalytic systems far from equilibrium may create ordered structures (e.g., catalytic cycles or active sites interacting in complex mechanisms), indicating that locally, these systems minimize entropy [9].

5.2.2   Topological Aspects of Catalytic Systems and Real Surfaces

Since several complex structures may occur during catalytic processes, it is often difficult to predict their catalytic behaviour. However, catalytic systems can be topologically described by polytopes, namely, geometrical structures with n dimensions in space. Table 5.1 reports the standard names of some n-polytopes.

Therefore, clusters, nanoparticles, enzymes and metal complexes in homogeneous media can be topologically described by 3-polytopes (polyhedra), whereas catalytic films (e.g., photoelectrocatalytic materials) can be described by 2-polytopes. Some examples are shown in Fig. 5.5.

We know that catalysts and biocatalysts are dynamic systems acting in a space–time domain. Thus, the kinetic and diffusive phenomena occurring on corners, edges and faces of ideal solid catalysts can be topologically described by time-related polytopes with $1(t)$, $2(s + t)$ and $3(2\,s + t)$ dimensions, respectively. However, real heterogeneous catalysts with surface roughness and structural defects can be better geometrically reflected by polytopes with $(3 + \varepsilon)$ dimensions, where ε represents the dimension of the boundary.

Table 5.1 Names of several n-polytopes

Dimension of polytope (n)	Name
1	Dion
2	Polygon
3	Polyhedron
4	Polychoron

Au147 gold cluster   Wilkinson's catalyst   Myoglobin   Photoelectrocatalytic film

Fig. 5.5 Examples of catalytic systems topologically described by 3-polytopes (Au 147 gold cluster, Wilkinson's catalyst and myoglobin) and a photoelectrocatalytic film topologically reflected by a 2-polytope

As shown in Table 5.2, several catalytic systems can be topologically described by 3-polytopes, whereas catalytic films are reflected by 2-polytopes. Such polytopes can be either convex or nonconvex.

Consequently, kinetic and diffusive phenomena occurring on real catalysts can be described by time-related polytopes as summarized in Table 5.3. Catalytic systems may lead to different molecular mechanisms, and such phenomena can be described by time-related polytopes that fuse the n dimensions of space and the one dimension of time into a single (n + 1)-dimensional manifold.

According to this classification, enzymes and homogeneous catalysts can be described by higher-dimensional polytopes (4-polytopes). Indeed, enzymatic reactions occur in the 3-D clefts of enzymes. Similarly, homogeneous catalysts (e.g., Wilkinson's catalyst) involve 3-D active sites distributed within homogenous media, and reactants directly undergo chemical binding to a specific catalytic group.

Although the molecular mechanisms of enzymes and homogeneous catalysts are expected to differ, in both cases, reactants easily interact with active sites that are fully accessible in a 3-dimensional space. Substantial differences can be observed in heterogeneous catalysts even if the latter can be reflected by 3-polytopes (see Table 5.2). In fact, reactants typically adsorb to the solid surface and then interact

Table 5.2 Topological properties of some catalytic systems

Catalyst	Geometrical dimensions	Topological properties
Heterogeneous catalyst (ideal)	3	3-polytope (either convex or nonconvex)
Heterogeneous catalyst (real)	$3 + \varepsilon$	$\approx$ 3-polytope (nonconvex)
Catalytic film	2	2-polytope (either convex or nonconvex)
Enzyme	3	3-polytope (nonconvex)
Homogeneous catalyst	3	3-polytope (nonconvex)

Table 5.3 Phenomenological properties of different catalytic systems

Catalyst	Molecular mechanisms	Space–time domain	Time-related polytopes
Heterogeneous catalyst (ideal)	Adsorb and React	$2s^a + t$	3-polytope
Heterogeneous catalyst (real)	Adsorb and React	$(2s + \varepsilon)^b + t$	$\approx$ 3-polytope
Catalytic film	Adsorb and React	$2s + t$	3-polytope
Enzyme	Adsorb and React	$3^c + t$	4-polytope
Homogeneous catalyst	React	$3^d + t$	4-polytope

[a]Ideal surface
[b]Real surface
[c]Cleft
[d]Single atom or group
ε the dimension of the boundary

with a collective electronic structure created by the presence of many atoms packed together. This confirms that both enzymes and homogeneous catalysts can be topologically more favoured in terms of outcomes than heterogeneous catalysts as their active sites are spatially more accessible.

Reactions on ideal solid surfaces occur in a 2-dimensional space. However, real surfaces can be highly complex due to the presence of various structural defects (e.g., terraces, kinks, adatoms, steps and vacancies). Moreover, real surfaces may exhibit asperities, roughness and fractal structures that play a key role in catalytic activity (Fig. 5.6).

The roughness of solid surfaces may promote the formation of local electric fields (E_L), which are beneficial for heterogeneous catalysis. In fact, local electric fields arise in dielectric and semiconductor-type materials with point defects that are electronically charged and can be intrinsic (thermally generated in a crystal) or extrinsic (impurity or dopant) [12].

Figure 5.7 shows the effect of the surface roughness on the electric field of solid catalysts. It appears that the electric field on flat surfaces (E_s) can be enhanced by the presence of structural defects. Therefore, the local electric field can be estimated as follows:

$$E_L \approx E_S \cdot \frac{l}{r}$$

where l and r correspond to the height and curvature radius of the structural defect, respectively.

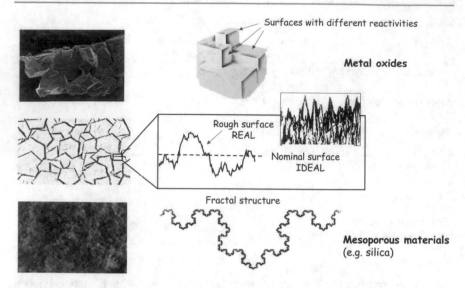

Fig. 5.6 Examples of real solid catalysts exhibiting different crystallographic planes, rough surfaces and fractal structures

Fig. 5.7 Local electric field (E_L) on a defective (rough) surface with electric field (E_s)

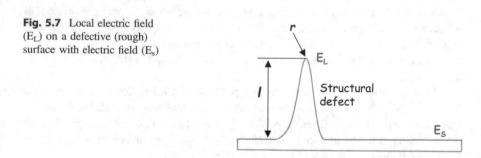

5.3 Well- and Ill-Conditioned Systems [9]

Stability is an important property of dynamical systems, indicating that their qualitative behaviour is not affected by small perturbations. In general, an equilibrium solution f_e to a system of first-order differential equations is stable if for every (small) $\epsilon > 0$, there exists a $\delta > 0$ such that for every solution $f(t)$ having initial conditions within distance δ

$$\|f(t_0) - f_e\| < \delta$$

the equilibrium remains within distance ϵ

$$\|f(t) - f_e\| < \epsilon$$

for all $t \geq t_0$.

In this scenario, the condition number k of a function with respect to an argument x measures the degree to which the output value of the function is affected by a small change in the input argument. Thus, a catalytic system with a low condition number is well conditioned, while a system with a high k value appears ill conditioned. Notably, the condition number is an intrinsic property of the system even if the operating conditions can modify the stability domain.

Thus, given a small change (Δ) in the x variable, the relative variation in x is

$$\frac{\Delta x}{x}$$

while the relative change in $f(x)$ is

$$\frac{f(x + \Delta x) - f(x)}{f(x)}$$

By comparing the ratios of the norms $\|\cdot\|$ for the domain/codomain of $f(x)$ and assuming an infinitesimal change δx, the following relation is obtained:

$$\frac{\|f(x + \delta x) - f(x)\|}{\|f(x)\|} \leq k \frac{\|\delta x\|}{\|x\|}$$

Thus, the system is well conditioned with $k \approx 1$.

The same approach can be applied to n variables $(x_1, x_2, \ldots, x_n \neq 0)$, where

$$f(x + \delta x) = f(x) + \delta x_1 \frac{\partial f(x)}{x_1} + \delta x_2 \frac{\partial f(x)}{x_2} + \ldots + \delta x_n \frac{\partial f(x)}{x_n} + O\|\delta x\|^2$$

Then, it is possible to obtain the following:

$$\frac{f(x + \delta x) - f(x)}{f(x)} = \frac{\delta x_1}{f(x)} \frac{\partial f(x)}{\partial x_1} + \frac{\delta x_2}{f(x)} \frac{\partial f(x)}{\partial x_2}$$

$$+ \ldots + \frac{\delta x_n}{f(x)} \frac{\partial f(x)}{\partial x_n} + O \frac{\delta x^2}{f(x)}$$

$$\leq \left|\frac{\delta x_1}{x_1}\right| \left|\frac{x_1}{f(x)}\right| \left|\frac{\partial f(x)}{\partial x_1}\right| + \left|\frac{\delta x_2}{x_2}\right| \left|\frac{x_2}{f(x)}\right| \left|\frac{\partial f(x)}{\partial x_2}\right| + \ldots + \left|\frac{\delta x_n}{x_n}\right| \left|\frac{x_n}{f(x)}\right| \left|\frac{\partial f(x)}{\partial x_n}\right|$$

$$= \left|\frac{\delta x_1}{x_1}\right| k_1 + \left|\frac{\delta x_2}{x_2}\right| k_2 + \ldots + \left|\frac{\delta x_n}{x_n}\right| k_n$$

The latter relation shows that greater k_n values occur when the derivative of the function $f(x)$ increases, i.e., when $f(x)$ exhibits a strong dependence on a specific variable, namely,

$$\frac{\partial f(x)}{\partial x_i} \gg 1.$$

Thus, if a catalytic process is highly sensitive to the initial conditions or small perturbations, it is an ill-conditioned system ($k \gg 1$) [9].

Several studies investigating N_2O decomposition over Cu-ZSM-5 catalysts revealed a reaction rate with singular and basically still unexplained oscillatory behaviour under specific operating conditions (e.g., temperature and residence time). More recently, it has been shown that both catalyst pretreatment and the nature of active species (Cu^{2+}/Cu^{+}) play a role in the oscillation pattern. A study combining IR spectroscopy and H_2-TPR conducted by Pirone et al. revealed the role of oligomeric Cu_xO_y species with extralattice oxygen close to the Al atoms in the oscillatory behaviour [7]. Figure 5.8 shows the N_2O outlet concentration values achieved during the isothermal reaction at 400 °C over the Cu-ZSM-5 catalysts prepared by impregnation, ion-exchange and sublimation routes. All catalysts exhibited certain activity in the decomposition of N_2O, although the level of activity varied among the samples and was dependent on both the Cu content and the nature of the Cu species. In particular, the catalysts prepared by impregnation (CZ-IM-A and CZ-IM-B) and the most Cu-rich exchanged zeolite sample (CZ-EX-B) showed significant and regular oscillatory behaviour. No oscillations were observed in CZ-EX-A due to its very low N_2O decomposition activity. Conversely, no oscillations were observed in the CZ-SU sample prepared by the sublimation procedure despite its significant activity. This finding is consistent with the fact that the CZ-SU sample was prepared with a Cu^{+} precursor, whereas both the impregnated and ion-exchanged catalysts were synthesized with a Cu^{2+} precursor. Moreover, it has been observed that the oscillation pattern varies with temperature; thus, the oscillation frequency increases with temperature (in the range 350–400 °C) [7].

The results show that this catalytic reaction reflects an ill-conditioned system in which activity strongly depends on the nature of the active sites and the operating conditions.

Another characteristic example of an ill-conditioned system is the CO oxidation reaction over ceria-based nanocatalysts. Several studies have shown that such a reaction may exhibit surface-sensitive behaviour, namely, small variations in the catalyst structure can produce large effects on oxidation activity. This dependency can be revealed by the corresponding Miller indices (h, k, l) or, equivalently, the unit vector normal to the plane (Fig. 5.9).

$$\mathbf{n} = (n_x, n_y, n_z) = \frac{(h, k, l)}{\sqrt{h^2 + k^2 + l^2}}$$

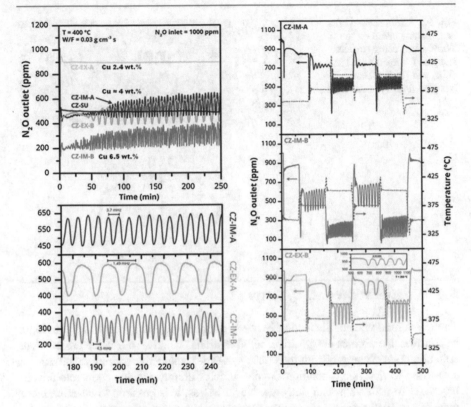

Fig. 5.8 Outlet concentration of N_2O as a function of time during catalytic activity testing under isothermal conditions (left). The results in terms of the N_2O outlet concentration (straight line) and reactor temperature (dashed line) as a function of time during testing with oscillation-inducing catalysts under different isothermal conditions (right). Reprinted and adapted from Armandi et al. [7] with permission from Elsevier

Crystalline solids spontaneously form particles with planes that exhibit the lowest surface energies (γ_{xyz}). The exposure of specific crystallographic facets and the presence of structural defects (e.g., edges or corners) are critical for controlling surface reactivity. Recent studies have shown that ceria-based nanomaterials with well-defined (100) and (110) planes are usually more active in oxidation reactions (e.g., CO oxidation and soot combustion) than polycrystalline ceria NPs with preferred exposure of (111) planes [10, 13, 14]. Therefore, it appears that structure-sensitive reactions are usually ill-conditioned systems since small variations in the catalyst surface properties may produce large effects on the catalytic activity [9].

Fig. 5.9 Cross section of a
nanoparticle based on the
Wulff shape using surface
energies (γ) for the (111),
(110) and (100) planes.
Adapted from Piumetti et al.
[15]

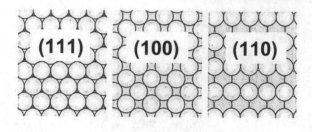

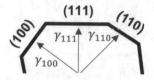

5.4   Cooperation and Synergy

There is a fundamental distinction between cooperative and synergistic behaviours
as follows: the former are positive collaborations, whereas the latter are efficient
collaborations. Synergy (from the Greek syn-ergos, meaning working together) is a
term used to describe a condition under which different entities cooperate advan-
tageously to achieve a final outcome. Nevertheless, it is possible to describe these
concepts in mathematical terms by set theory and measure theory.

Let A and B be two sets with $p(A) \neq 0$, $p(B) \neq 0$ and $A \cap B \neq 0$. Then,
assume that A and B represent sets of actions. We define cooperation as the following condition:

$$p(A \cap B) > 0$$

We define synergy as the following condition:

$$p(A \, \Delta \, B) > 0$$

where $A \, \Delta \, B$ is the symmetric difference of sets A and B, namely,

$$A \, \Delta \, B = (A - B) \cup (B - A) = (A \cup B) - (A \cap B)$$

This means that synergy occurs when the set of elements that are in either of the
sets but not in their intersection is empty. A schematic of the concepts of cooperation and synergy is shown in Fig. 5.10.

During a catalytic reaction, either cooperative or synergistic phenomena may
occur as a result of the nonlinear combination of complex structures [16].

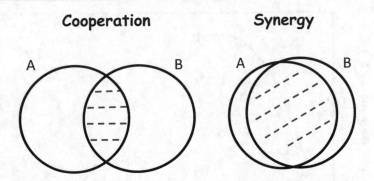

Fig. 5.10 Schematic of the concepts of cooperation and synergy

The benefits in heterogeneous catalysis typically arise through phase cooperation and spillover effects. Grasselli et al. [17] conducted phase-cooperation studies to investigate oxidation and ammoxidation reactions. The results showed that both the $\alpha\text{-}Bi_2Mo_3O_{12}$ and $\gamma\text{-}Bi_2MoO_6$ phases may exhibit promising redox activities by complementary physicochemical properties. Similarly, X-ray photoelectron spectroscopy (XPS) and electrical conductivity measurements have shown that multicomponent materials, such as bismuth molybdates over CoFe molybdates, show significantly improved redox-type activity because of the higher electrical conductivity (of the CoFe molybdate phase) attributable to both Fe^{2+} and Fe^{3+} species in Co^{2+} molybdates [16].

Multiphase cooperation and synergistic effects can be evaluated as follows:

$$\Delta r = r_{real-i} - \sum_{i=1}^{n} x_i \cdot r_i$$

where r_{real-i} denotes the real rates of n phases, and x_i and r_i denote the molar fractions and the theoretical (estimated) rates of the i-th phase, respectively.

For example, Piumetti et al. [18] showed the presence of cooperative-synergistic effects with a set of cerium-copper oxides (with different Cu/Ce values) in the oxidation of CO and ethylene (Fig. 5.11). The best overall performance of both oxidation reactions was achieved with catalysts with Ce/Cu ranging from 0.67 to 1.5. This finding was attributed to the presence of CeO_x and CuO_x domains that cooperate synergistically, thereby leading to higher activity because of the easier surface reducibility and more abundant structural defects.

In this case, phase cooperation can be reflected by the difference between the real specific oxidation rates (r_{real-i}) and the theoretical rates (r_i) estimated from a linear combination of rates and Cu/Ce molar fractions:

$$\Delta r = r_{real-i} - r_i$$

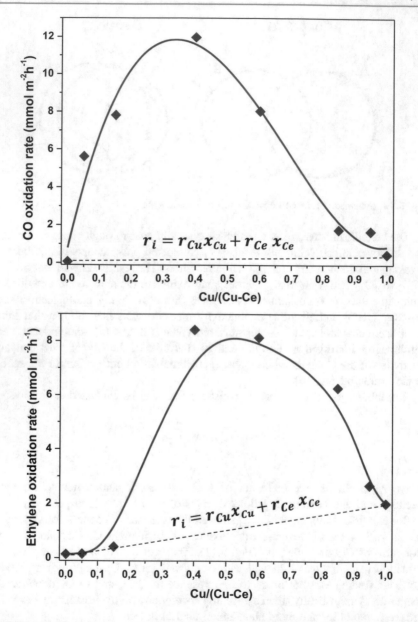

Fig. 5.11 Specific oxidation rates achieved over a set of cerium-copper oxide catalysts under isothermal conditions. CO oxidation reaction: catalyst = 0.1 g; flow rate = 50 ml min^{-1}; 1000 ppm CO and 10% O$_2$ in N$_2$; temperature = 90 °C. Ethylene oxidation reaction: catalyst = 0.1 g; flow rate = 100 ml min^{-1}; 500 ppm ethylene and 10% O$_2$ in N$_2$; temperature = 310 °C. Adapted from Piumetti et al. [18]

In which

$$r_i = r_{Cu} \cdot x_{Cu} + r_{Ce} \cdot x_{Ce}$$

That is, the theoretical reaction rate (r_i) expected for a specific molar fraction composition of the catalyst $(\sum x_i = 1)$.

Thus, we can introduce the interaction quotient (I) to quantify the beneficial effects resulting from interactions among active phases of multicomponent catalysts, namely:

$$I = \frac{r_{real-i}}{r_i}$$

where the theoretical reaction rates in multicomponent catalysts (with n-active phases) can be expressed as

$$r_i = r_\alpha \cdot x_\alpha + r_\beta \cdot x_\beta + \ldots + r_n \cdot x_n$$

Therefore, when $I > 1$, there are beneficial effects on the catalytic performance as a consequence of multicomponent interactions, whereas when $I = 1$, there are no significant interactions, and the whole activity results from a linear combination of n-rates and molar fractions. In contrast, when $I < 1$, inhibition effects occur in the catalytic system.

Delmon et al. [19] introduced the concept of remote control to explain why many industrial catalysts used for the oxidation of hydrocarbons are multiphasic and specific phase compositions lead to synergistic phenomena. Spillover phenomena at solid surfaces involve the transport of active species (e.g., oxygen, hydrogen, etc.) adsorbed on one phase (donor) onto a second phase (acceptor), which does not form active species under the same conditions [16].

As shown in Fig. 5.12, the "irrigation" of catalytic sites (M) by spillover oxygen species (e.g., O, O^-, O^{2-}, and O_2^{2-}) enhances the chemical reactivity of the catalytic surface towards soot combustion, thereby multiplying the oxidation activity of solid–solid catalytic systems mediated by gas-phase oxygen [20].

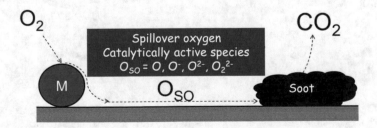

Fig. 5.12 Schematization of spillover oxygen species generated over catalytic surfaces active towards soot combustion. Adapted from Piumetti and Lygeros [20]

Cooperative and synergistic effects may also occur in homogeneous catalysis. Recent studies by Piumetti et al. [16] showed that the presence of both H_2O_2 and ascorbic acid in V-containing solutions strongly enhances the degradation of organic molecules via Fenton-type mechanisms. As shown in Fig. 5.13a, different combinations of vanadium (V), ascorbic acid (HA) and hydrogen peroxide (H_2O_2) were considered in a solution of azo dye acid orange-7 (AO7 = 10^{-3} M). In this scenario, the lowest A07 degradation values (%) after 30 min were obtained in systems with only one active component or player (Fig. 5.13b); the presence of either vanadium or hydrogen peroxide had a moderate effect on the degradation of A07 (approximately 5% degradation after 30 min), whereas worse activity was observed with ascorbic acid (1.8%). Conversely, the presence of two players promoted A07 conversion, and the best results were obtained with the HA/H_2O_2 system (AO7 degradation = 52.7%). Remarkable results (AO7 degradation = 100%) were achieved in a system with three players (V/HA/H_2O_2), thereby confirming the beneficial effects of both oxidant and reductant agents in promoting Fenton-type processes [16].

Similar phenomena can be observed in enzyme catalysis by allosteric effects. As previously reported (*vide* Chap. 2), allosteric activation occurs when the binding of one molecule enhances the affinity between substrate molecules and other binding sites. Thus, the binding of oxygen (substrate and effector) to haemoglobin induces a local conformational change such that the remaining active sites increase their affinity for oxygen (Fig. 5.14).

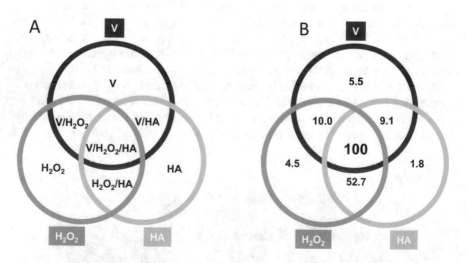

Fig. 5.13 Venn diagrams showing combinations of vanadium, ascorbic acid (HA) and hydrogen peroxide (H_2O_2) in a solution of acid orange-7 (AO7) and the initial concentrations of V = 2×10^{-2} M (NH_4VO_3) HA = 8×10^{-2} M and H_2O_2 = 8×10^{-2} M (Section A). AO7 conversion values (%) of different homogeneous catalytic systems achieved after 30 min (Section B). Adapted from Piumetti and Lygeros [16]

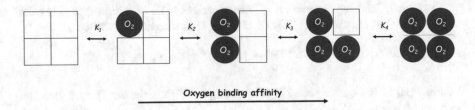

Oxygen binding affinity

Fig. 5.14 Cooperative binding of haemoglobin (the binding of oxygen facilitates the binding of more oxygen)

Since haemoglobin (Hb) has four binding sites for oxygen (haem), its equilibrium with oxygen can be defined by the following equation:

$$Hb(O_2)_{n-1} Hb(O_2)_n$$

where n = 1,2,3,4. Thus, there are four corresponding equilibrium constants K_n (Adair constants), and the sigmoid oxygen equilibrium curve indicates that $K_1 < K_4$. This phenomenon indicates that the binding reactions at individual sites in each haemoglobin molecule are not independent of one another.

Entropy plays a key role in this cooperativity. Indeed, the first oxygen can bind one of four haem groups. This corresponds to a higher entropy than the binding of the last oxygen, which has only one haem left available to bind. This entropy difference may explain the positive cooperativity in oxygen binding haemoglobin.

5.5 New Modelling Approaches for Decoding Complexity in Catalysis

5.5.1 The Theory of Hyperstructures

Introduced in 1934 with hypergroups by Marty [21], the theory of hyperstructures represents an extension of the classical algebraic structures. In fact, in an algebraic structure, the composition of elements is still an element, while in an algebraic hyperstructure, the composition of elements is a set [20].

By definition, a hyperstructure is every algebraic structure in which at least one hyperoperation (multivalued operation) can be defined.

A classical operation (·) on a set H is any map from H × H. This means that for any two elements x, y ∈ H, there is an element of H corresponding to x · y. This map can be written as

$$\cdot : H \times H \to H : (x, y) \to x \cdot y \in H$$

Addition (+) and multiplication (·) are the typical operations.

Conversely, the hyperoperation (O) in a set H is the operation that maps the elements x, y of H to a subset $x\text{O}y$ of H [22]. This map can be written as

$$\circ : H \times H \rightarrow P(H) - \{\emptyset\} : (x, y) \rightarrow x \circ y \in H$$

where $P(H)$ is the power set of H. For example, if H is the set $\{x, y,\}$ then the subsets of H are

$$\{\emptyset\},$$
$$\{x\},$$
$$\{y\},$$
$$\{x, y\},$$

and hence the power set of H is the set of all subsets of H, including the empty set, namely,

the set of all functions from H to a given set of two elements, 2^H.

Moreover, a hyperstructure (H, O) is a hypergroup if (O) is an associative hyperoperation for which the reproduction axiom

$$iH = Hi = H, \forall i \in H$$

is valid according to Marty's definition [21].

In these terms, the catalyst acts as a hyperoperator (O), computing a nonlinear combination of different complex structures (C_i). Thus, the complexity of a catalytic reaction (λ) can be written as

$$\lambda = C_1 \circ C_2 \circ C_3 \ldots \circ C_i$$

with the power set of 2^{C_i}. This approach can be used to describe complex catalytic behaviours and predict the reaction products [9].

In this scenario, T. Vougiouklis introduced Hv-structures, the largest classes of hyperstructures, which have been enumerated by Lygeros [23–25]. Hv-structures have attracted attention in several research areas, including linguistics, biology, chemistry, and physics. The fundamental concept for these Hv groups is the weak associativity. The hyperoperation is weak associative (WASS) if

$$(xy)z \cap x(yz) \neq \emptyset, \forall x, y, z \in H$$

Thus, the two sets $(xy)z$ and $x(yz)$ have at least one common element for all x, y, z in H [21, 23]. The hyperstructure (H, O) is called H_v-semigroup if it is WASS, whereas it can be named H_v-group if it is a reproductive H_v-semigroup according to the reproduction axiom, e.g., $xH = Hx = H \forall x \in H$ [24].

Hv-structures can be useful to describe the complexity of catalysts. In fact, the complex structures of active sites are Hv-structures, the largest class of hyper-structures that satisfy the weak properties [10, 20]. During a reaction, the catalyst acts as a hyperoperator, mapping the reactants to the class of time-related polytopes (vide supra), which is a subset of H_v-structures. Since time-related polytopes reflect catalytic systems that vary over time, they represent the initial set of actions of the hyperstructures (Hv-structures). The hyperoperation is WASS because different complex structures of active sites (subsets) contain time-related polytopes as common elements. Thus, the condition required for the existence of the Hv-structures is satisfied. Finally, the reaction products belong to a set that is usually not a singleton [10, 20].

5.5.1.1   Cooperation and Synergetic Hyperstructures [16]

In algebraic structures, a group G can be defined with the necessary but not suf-ficient condition

$$\forall(x,y) \in G^2 : |x \cdot y| = 1$$

where x and y are two elements of G. Moreover, in G, there is an inner operation ($\cdot$), which means that

$$(x,y) \in G^2 : |x \cdot y| = G$$

Therefore, the elements x and y may cooperate to give a result (or product). Conversely, in a hypergroup H, the following condition has to be satisfied:

$$\forall(x,y) \in H^2 : 1 \leq |x \circ y| \leq |H|$$

which is a consequence of the axioms of reproduction and associativity (vide supra).

Thus, it is possible to have a formal distinction between cooperation and synergy via the theory of hyperstructures. In fact, there is cooperation when the group operation is

$$|x \circ y| = 1$$

and there is synergy when the following condition holds:

$$|x \circ y| > 1$$

Then, a synergetic hyperstructure (SH) is defined as

$$\forall(x,y) \in SH : |x \circ y| > 1$$

and a strong synergetic hyperstructure (SSH) is defined as

$$\forall (x, y) \in SSH : |x \circ y| > 2$$

Therefore, strong synergy among the elements occurs when

$$x \cup y = \{x, y\} \subset x \circ y$$

These theoretical concepts should be applied to catalysis, leading to a more formal classification of cooperation and synergy in catalysis. As shown in Table 2, the classification is based on the number of reaction products. Therefore, a scale of magnitude in terms of catalytic benefits can be proposed as follows:

Cooperation < Synergetic Hyperstructure < < Strong Synergetic Hyperstructure

Therefore, SSH-type catalytic systems may exhibit greater catalytic improvements than SH- and C-type systems (Table 5.4).

According to this classification, the catalytic oxidation of CO

$$CO + {}^{1}\!/_{2}O_2 \rightarrow CO_2$$

is a characteristic example in which cooperation is possible, whereas the catalytic oxidation of volatile oxygen compounds (VOCs) (reaction products = CO_2 and H_2O)

$$2H_xC_y + (2y + {}^{1}\!/_{2}x)O_2 \rightarrow 2yCO_2 + xH_2O$$

and the catalytic decomposition of N_2O (reaction products = N_2 and O_2)

$$N_2O \rightarrow N_2 + {}^{1}\!/_{2}\, O_2$$

are both examples of catalytic systems in which SHs may occur. On the other hand, catalytic reactions leading to n distinct products (for $n \geq 3$) such as the total oxidation of chlorinated volatile organic compounds

$$2H_xC_yCl_z + (2y + {}^{1}\!/_{2}\, x)O_2 \rightarrow 2yCO_2 + xH_2O + zCl_2$$

can achieve strong SHs. A similar classification can also be applied to homogeneous and enzyme catalysis.

Table 5.4 Classification of catalytic systems based on the number of products

Classification	Reaction products
Cooperation (C)	1
Synergetic hyperstructure (SH)	2
Strong synergetic hyperstructure (SSH)	≥ 3

5.5.1.2   Strategic Relevance [16]

The role of strategic relevance in terms of benefits must be noted. As defined by Hjørland Sejer Christensen [26], something is relevant to a task if it rises the likelihood of accomplishing the goal implied by the task. Thus, relevance refers to the task, but the task must be codified as a strategic mix to be global and not only local. In fact, relevance is related to the efficiency of contribution in this context. If the contribution of some elements in the strategic mix is negative (e.g., impurities), it is not relevant. In contrast, relevance is a positive contribution to the strategic mix (strategic relevance), which is related to the concept of cooperation. Moreover, strategic relevance can be strong in the context of synergistic behaviour (efficient collaboration). However, relevance can be passive and, hence, not active in a strategic mix. Thus, the concept of relevance, specifically strategic relevance, is not independent of the task and rather depends on the strategic mix. Thus, strategic relevance is a condition necessary for catalytic systems to obtain a specific goal [16].

5.5.2   Application of Game Theory to Catalytic Systems [16]

The game theory approach provides tools for analysing the conditions under which parties (players) make rational, interdependent decisions. A solution to a game describes the optimal decisions of the players and the outcomes that may arise from these decision-makers.

For each game G (strategic form games), we have to define

- The set of players (P)
- The strategies (S)
- The payoff (utility) functions (F)

More formally:
A strategic form game is a triplet

$$G = \langle P, (S_i)_{i \in P}, (F_i)_{i \in P} \rangle$$

where

P is a finite set of players, e.g., $P = \{1, ..., l\}$;

S_i is the set of available actions for player i;

$s_i \in S_i$ is an action for player i;

$F_i: S \to R$ is the payoff (utility) function of player i, where $S = \prod_i S_i$ is the set of all action profiles.

To apply game theory to chemical systems, we must assume that the chemical choice is rational. Thus, the players (catalytically active sites or components) exhibit a number of pure strategies as reflected by their concentrations and play a strategic game.

A characteristic example of synergy among catalytically active components is the homogeneous catalytic system effective for the degradation of the azo dye acid orange-7 (vide supra). In this case, AO7 degradation can be carried out using V species in the homogeneous phase along with H_2O_2 and HA as oxidant/reductant agents, respectively [16]. In the normal form, this catalytic system has the following structure:

$$G = \langle P, S, F \rangle$$

where

$$P = \{V, H_2O_2, HA\}$$

is the set of players, and

$$S = \left\{S_V, S_{H_2O_2}, S_{HA}\right\}$$

is a 3-tuple of pure strategy sets, one per player. Similarly, a 3-tuple of payoff functions can be

$$F = \left\{F_V, F_{H_2O_2}, F_{HA}\right\}$$

AO7 degradation (%) can be assumed to be a payoff function

$$AO7degr.(\%) = \frac{C_0 - C_t}{C_o} \cdot 100$$

where C_o and C_t represent the AO7 concentration under the initial conditions (10^{-3} M) and after a defined reaction time, respectively. As shown in Fig. 5.15, seven combinations of players (active components) with specific strategies (concentrations) play a strategic game in an AO7 solution as follows:

- V, H_2O_2 and HA (three single players);
- V/H_2O_2, V/HA and H_2O_2/HA (three pairs of players);
- V/HA/H_2O_2 (a triple of players).

Under these conditions, the players have a finite number of pure strategies as reflected by their concentrations, and they play strategic games. When the players act simultaneously, a synchronous game occurs. Moreover, since the outcome (AO7 degradation) is greater than zero, it is a non-zero-sum game. According to game theory, a non-zero-sum game is a situation in which the interacting parties' aggregate gains and losses can be more (or less) than zero. Moreover, there is a

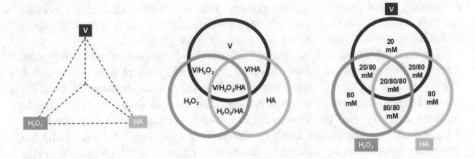

Fig. 5.15 Combinations of players (V = vanadium; HA = ascorbic acid and H_2O_2 = hydrogen peroxide) and their strategies (concentrations) acting in a homogeneous solution of acid orange-7. Adapted from Piumetti and Lygeros [16]

Nash equilibrium (NE) since each player performs the best action in a cooperative game while considering the actions of the other players. Each strategy adopted by the players in NE is the best response to all other possible strategies for that equilibrium.

As previously reported (Fig. 5.13), the lowest AO7 degradation values (%) appear in systems with a single player in solution (e.g., V, HA or H_2O_2). Each player acts individually, and thus, its catalytic action is a pure strategy. The payoff function of each player can be ordered in a set of pure strategies according to the AO7 degradation values (%) achieved in each catalytic system:

$$F_{HA} < F_{H_2O_2} < F_V (\text{pure strategy})$$

In the presence of two players, AO7 degradation (%) increases, and the best performance can be obtained with the HA/H_2O_2 pair (AO7 degr.% = 52.7 after 30 min). Since two players (V/H_2O_2, V/HA or H_2O_2/HA) act together, a mixed strategy is implemented. Thus, the following order of payoff functions in a set of mixed strategies can be described:

$$F_{V/HA} < F_{V/2O_2} < F_{H_2O_2/HA} (\text{mixed strategy})$$

As observed by Nash, if each player has a finite number of pure strategies, there exists at least one equilibrium in mixed strategies. Therefore, the catalytic system reaches intrinsic equilibrium during the catalytic process. Finally, remarkable results occur in AO7 degradation (%) when three players are simultaneously present in solution, thus leading to the total degradation of A07 within 10 min. The players act together with a fully mixed strategy, and the payoff function is defined as

$$F_{V/H_2O_2/HA} (\text{totally mixed strategy}).$$

Overall, these catalytic systems remain under a condition of Pareto efficiency because it is not possible to improve any payoff (utility) function without changing the other conditions (e.g., the set of players or strategies).

The same approach can be used to describe the behaviour of multicomponent catalysts. For example, a mixed oxide catalyst should have two (or more) active phases (α and β players) interacting with each other, and the magnitude of the payoff function depends on the α to β ratios. This catalyst should activate a mixed strategy to carry out the strategic game under a Pareto efficiency condition. Moreover, depending on the nature of the catalytic system, the strategic game can be either cooperative or synergetic, leading to different catalytic benefits (vide supra). As a result, game theory seems to be a useful mathematical tool for describing the dynamics of complex catalytic systems [16].

5.6 Artificial Intelligence Faces Catalytic Complexity

Combinatorial catalysis

In the early nineteenth century, ammonia was synthesized at the industrial level for the first-time using iron-based catalysts. More than 6500 experiments were performed, and approximately 2500 different catalysts were tested before reaching the large-scale production of NH_3 (Bosh-Haber process). This process was a true example of the Edisonian approach, which required a significant labour force to work in a systematic way over a long period.

The recent need for the fast screening of catalyst compositions motivated many scientists and engineers to develop combinatorial catalysis analogous to the combinatorial methods used for drug discovery. For example, Reichenbach et al. [27] introduced a library of catalytic materials prepared by organic precursors dispersed in metal nitrate solutions (polymerizable-complex-method). Thus, the drop-on-demand printer can synthesize catalysts with different compositions by combining syringe pumps with inkjet dispensers and incorporating a PC-controlled table acting in 3-D (Fig. 5.16).

This automatized catalyst synthesis approach has also been extended to testing, analysis and data handling. However, this research field is still in its infancy, and whether combinatorial catalysis could be more efficient in discovering new formulations than conventional trial and error approaches remains unclear. The most important aspect in combinatorial catalysis is probably highly parallelized testing, which provides activity data related to the performance of catalysts under different operating conditions. Similarly, some characterization techniques for activity screening have been proposed, including IR thermography, mass spectrometry, fluorescence detection and multichannel reactors with multiport valves for switching to analytical instruments [28].

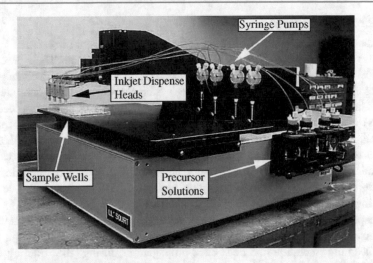

Fig. 5.16 Inkjet dispensing printer used for the synthesis of catalysts. Reprinted from Reichenbach et al. [27] with permission from Elsevier

Statistical design of experiments (DOE) and optimization methods

The use of computer-assisted methods for the design of experiments depends on whether the catalysts have to be investigated systematically and whether the goal is to search for the catalysts with the best performances. In fact, the former case is a task for the statistical design of experiments (DOE), whereas optimization methods are more appropriate for the latter.

Statistical DOE was introduced in the 1970s. These methods are typically used to solve more ambitious and computationally much more demanding tasks. Among the computer-aided DOE methods available, D-optimal design is probably the most frequently used. It consists of a model assuming a dependent variable (e.g., yield, etc.) that depends linearly on the input variables and their interactions.

However, when it is not required (or not possible) to investigate the entire set of catalytic materials but only one or several small clusters of highly performing materials, *optimization methods* are more appropriate. Indeed, high catalytic performance is typically achieved only in small ranges of compositions, and such methods allow us to search for optimum functions describing the dependency of catalyst performance on their composition. Although the latter must be obtained empirically (i.e., through experimental measurements), functions exist (i.e., those describing the reaction kinetics) that may be obtained analytically [29]. Among these functions, genetic algorithms functioning with a heuristic approach are the most frequently used. However, these methods often require preliminary knowledge of the data for correct tuning. Therefore, genetic algorithms are frequently combined with other methods, e.g., neural networks.

Artificial neural networks

Advances in artificial intelligence have led to the development of machine learning
and deep learning, which attempt to mimic human intelligence. These new technological advances may contribute to data mining in catalysis and biocatalysis.

The most recent artificial intelligence systems are based on deep learning
models, a subset of machine learning methods centred on artificial neural networks
(ANNs) with representation learning. Deep learning belongs to a class of machine
learning algorithms that uses multiple layers to progressively extract higher-level
features from the data (input layer). The most often used ANN for catalytic
applications is the so-called multilayer perceptron (MLP), which provides fairly
accurate solutions for extremely complex problems. MLP consists of at least the
following three layers of nodes as illustrated in Fig. 5.17.

Except for the input nodes, each node is a neuron that uses nonlinear activation
functions (e.g., sigmoid, rectifier, and hyperbolic tangent). MLP utilizes a super-
vised learning technique called back-propagation during the training phase
(Fig. 5.18). Thus, learning occurs in the perceptron by varying the associated
weights (w_i) after each data point is processed based on the amount of error in the
output compared with target values.

Therefore, when an ANN is used in a catalyst development program, it is
necessary to "train" the network using real values from a training library with
known data. After verification of the training with another set of materials with
known performance, the network can be used to predict the performance of catalysts (catalytic screening) [29].

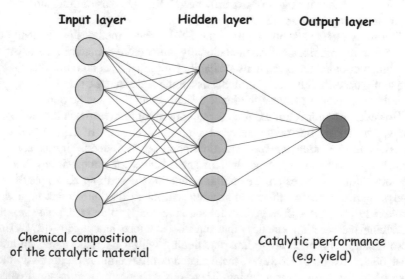

Fig. 5.17 Example of an artificial neural network architecture with one layer of hidden neurons
employed to approximate the unknown performance of a heterogeneous catalytic reaction as a
function of the catalyst composition. Adapted from Ertl et al. [29]

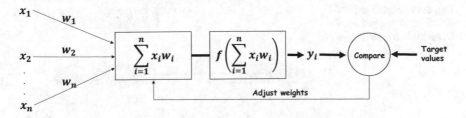

Fig. 5.18 Building block of ANN with back-propagation. Each input x_i has an associated weight w_i. The sum of all weighted inputs, $x_i w_i$, is passed through a nonlinear activation function f to give an output y_i. The latter can be compared with target values, and then, a back-propagation algorithm allows the weights to be adjusted

ANNs have proven to be valuable tools for biochemical data modelling as they are known to be effective function approximators. Specifically, when designed and trained properly, ANNs may approximate continuous functions to any desired accuracy level. Thus, ANNs can be useful for addressing some fundamental problems related to the folding process and the structure–function relationship in proteins. In particular, deep neural networks have attracted considerable interest in recent years and can currently predict the 3-D structure of a protein based solely on its amino acid sequence. Among these models, the AlphaFold system developed by the team of DeepMind (founded in London in 2010) appears to be a considerable advance in protein-structure prediction. The 3-D models of proteins generated by AlphaFold are particularly accurate, even with sequences with fewer homologous sequences [30]. As shown in Fig. 5.19, the AlphaFold algorithm works with a procedure called "multiple sequence alignment", which consists of a comparison of the sequence of a target protein with similar sequences in the database. Therefore, AlphaFold predicts the distance between two amino acids responsible for the target protein via the exact distances of proteins with a similar sequence present in the database. A second neural network performs similar work for neighbouring amino acid binding angles in the primary chain. The predicted distances or angles are related to a score that estimates the accuracy of the structure. Then, AlphaFold creates a physically possible structure and, using an iterative mathematical method (gradient descent), refines the structure to obtain the previously predicted distances and angles.

Different approaches may be used to find specific compounds that can be developed and marketed. The basic research required for pharmaceutical development dictates that 10,000–30,000 substances must undergo laboratory screening for each new drug approved for use in humans. As a result, the overall process from the discovery to the post-marketing of a drug can take 10–15 years. Recent studies have estimated that new drugs cost US \$2.6 billion to research and develop, including the price of failure and opportunity costs [32]. In this scenario, the use of artificial intelligence to predict stable 3-D structures of proteins appears to be a sustainable approach to better understanding physiological processes and diseases.

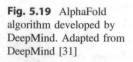
Fig. 5.19 AlphaFold
algorithm developed by
DeepMind. Adapted from
DeepMind [31]

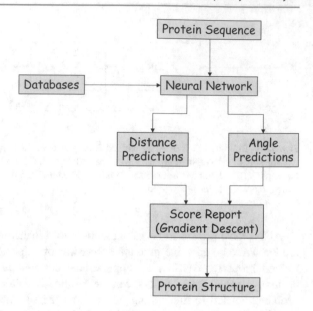

Moreover, this approach could improve the ability to design new targeted bio-therapeutics in the shortest possible time. However, the recent pandemic situation caused by COVID-19 illustrated that both local and global efforts can significantly shorten vaccine development timelines.

5.7   Summary

- Self-organizing phenomena in catalysis reflect the presence of complex structures in dynamic systems. These complex (ordered) structures occurring in heterogeneous, homogeneous and enzyme catalysis may result in different microscopic, mesoscopic and macroscopic phenomena.
- Catalytic systems can be topologically reflected by polytopes.
- Catalytic reactions can lead to either ill- or well-conditioned systems. When the catalytic process is highly sensitive to the initial conditions or small perturbations, it is an ill-conditioned system.
- During a catalytic reaction, either cooperative or synergistic phenomena may occur as a result of the nonlinear combination of complex structures.
- Game theory and hyperstructure theory may describe the behaviour of catalytic systems fairly well. Accordingly, it is possible to describe the dynamics of complex catalytic systems and correctly and simultaneously compute the reaction products.

- Progress in artificial intelligence has led to the development of machine learning and deep learning that mimic human intelligence. These new technological advances are beneficial for progress in research related to catalysis and biocatalysis.

5.8 Questions

1. Self-organizing phenomena in catalysis reflect the presence of complex structures acting in dynamic systems. Provide some examples of these structures occurring in heterogeneous, homogeneous and enzyme catalysis.
2. Why are enzymes and homogeneous catalysts topologically more favoured in terms of outcomes than real solid catalysts? What are the corresponding time-related polytopes of these catalytic systems?
3. Provide some examples of ill- and well-conditioned systems of catalysts.
4. Overall, there is a fundamental distinction between cooperative and synergistic catalytic behaviours. Prove this distinction using i) set theory and ii) hyperstructure theory.
5. What are the main benefits of applying artificial intelligence to catalysis and biocatalysis? What could be the future perspectives of AI in these research areas?

References

1. M. Piumetti, N. Lygeros, Chimica Oggi—Chem. Today **31**, 48–52 (2013)
2. G. Ertl, *Reactions at Solid Surfaces* (Wiley-VCH, Hoboken, 2009)
3. G. Nicolis, I. Prigogine, *Self-Organization in Non-Equilibrium Systems* (Wiley-VCH, New York, 1977)
4. I. Prigogine, Ann. N.Y. Acad. Sci. **988**, 128–132 (2003)
5. G. Ertl, Angew. Chem. Int. Ed. Engl. **29**(11), 1219–1227 (1990)
6. S. Hawking S., R. Penrose, *The Nature of Space and Time* (Princeton University Press, Princeton, 1996)
7. M. Armandi, T. Andana, S. Bensaid, M. Piumetti, B. Bonelli, R. Pirone, Catal. Today **345**, 59–70 (2020)
8. G. Nicolis, C. Nicolis, *Foundations of Complex Systems* (World Scientific Publishing, 2007), pp. 247–274
9. M. Piumetti, N. Lygeros, Chimica Oggi—Chem. Today **34**(5), 6–10 (2016)
10. M. Piumetti, N. Lygeros, Hadronic J. **36**, 177–195 (2013)
11. M.C. Yovits, S. Cameron, *Self-Organizing Systems* (Pergamon Press, London, 1960), pp. 31–50
12. D.E. Milovzorov, in *Electric Field*, ed. by M.S. Kandelousi (Intechopen, 2018)
13. M. Piumetti, T. Andana, S. Bensaid, D. Fino, N. Russo, R. Pirone, AIChE J. **63**(1), 216–225 (2017)
14. S. Bensaid, M. Piumetti, C. Novara, F. Giorgis, A. Chiodoni, N. Russo, D. Fino, Nanoscale Res. Lett. **11**(494), 1–14 (2016)
15. M. Piumetti, S. Bensaid, D. Fino, N. Russo, Appl. Catal. B **197**, 35–46 (2016)

16. M. Piumetti, N. Lygeros, Chimica Oggi—Chem. Today **33**(6), 38–44 (2015)
17. R.K. Grasselli, Top. Catal. **15**, 93–101 (2001)
18. M. Piumetti, S. Bensaid, T. Andana, N. Russo, R. Pirone, D. Fino, Appl. Catal. B **205**, 455–468 (2017)
19. B. Delmon, Heterog. Chem. Rev. **1**, 219–230 (1994)
20. M. Piumetti, N. Lygeros, Chimica Oggi—Chem. Today **33**(1), 46–49 (2015)
21. F. Marty, *Sur une généralisation de la notion de groupe* (8th Congress Math. Stockholm, 1934), pp. 45–49
22. T. Vougiouklis, *New Frontiers in Hyperstructures* (Hadronic Press, Palm Harbor, 1996), p. 48
23. T. Vougiouklis, Hyperstructures and Their Representations (Hadronic Press Inc., Florida, 1994); B. Davvaz, T. Vougiouklis, *A Walk Through Weak Hyperstructures: Hv-Structures* (World Scientific, 2018), p. 348
24. N. Lygeros, T. Vougiouklis, Ratio Math. **25**, 59–66 (2013)
25. R. Bayon, N. Lygeros, J. Algebra **320**(2), 821–835 (2008)
26. B. Hjørland, F. Sejer Christensen, JASIST **53**(11), 960–965 (2002)
27. H.M. Reichenbach, A. Hongmei, P.J. McGinn, Appl. Catal. B **44**, 347–354 (2003)
28. B. Cornils, W.A. Herrmann, M. Muhler, C.-H. Wong, *Catalysis from A to Z* (Wiley-VCH, Weinheim, 2007), p. 334
29. G. Ertl, H. Knözinger, F. Schüth, J. Weitkamp, *Handbook of Heterogeneous Catalysis*, 2nd ed. (Wiley-VCH, Weinheim, 2008), pp. 66–81 and 2053–2074
30. A.W. Senior, R. Evans, J. Jumper, J. Kirkpatrick, L. Sifre, T. Green, C. Qin, A. Žídek, A.W. R. Nelson, A. Bridgland, H. Penedones, S. Petersen, K. Simonyan, S. Crossan, P. Kohli, D.T. Jones, D. Silver, K. Kavukcuoglu, D. Hassabis, Nature **577**, 706–710 (2020)
31. DeepMind, https://deepmind.com
32. A. Mullard, Nat. Rev. Drug Discovery **13**, 877 (2014)

Printed in the United States
by Baker & Taylor Publisher Services